Reviews of Environmental Contamination and Toxicology

VOLUME 111

Reviews of Environmental Contamination and Toxicology

Continuation of Residue Reviews

Editor
George W. Ware

Editorial Board
F. Bro-Rasmussen, Lyngby, Denmark
D.G. Crosby, Davis, California · G.H. Hudson, Overijse, Belgium
H. Frehse, Leverkusen-Bayerwerk, Germany
H.F. Linskens, Nijmegen, The Netherlands
O. Hutzinger, Bayreuth, Germany · N.N. Melnikov, Moscow, U.S.S.R.
M.L. Leng, Midland, Michigan · R. Mestres, Montpellier, France
D.P. Morgan, Oakdale, Iowa · P. De Pietri-Tonelli, Milano, Italy
Raymond S.H. Yang, Research Triangle Park, North Carolina

Founding Editor
Francis A. Gunther

VOLUME 111

Springer-Verlag
New York Berlin Heidelberg
London Paris Tokyo Hong Kong

Coordinating Board of Editors

GEORGE W. WARE, *Editor*
Reviews of Environmental Contamination and Toxicology

College of Agriculture
University of Arizona
Tucson, Arizona 85721, USA
(602) 621-7201

HERBERT N. NIGG, *Editor*
Bulletin of Environmental Contamination and Toxicology

Agricultural Research and Education Center
University of Florida
700 Experimental Station Road
Lake Alfred, Florida 33850, USA
(813) 956-1151

ARTHUR BEVENUE, *Editor*
Archives of Environmental Contamination and Toxicology

35 Fifteenth Avenue
San Mateo, California 94402, USA
(415) 572-1029

New York: 175 Fifth Avenue, New York, N.Y. 10010, USA
Heidelberg: 6900 Heidelberg 1, Postfach 105 280, West Germany

Library of Congress Catalog Card Number 62-18595.
Printed in the United States of America.

ISSN 0179-5953

© 1990 by Springer-Verlag New York Inc.
All rights reserved. This work may not be translated or copied in whole or in part without the written permission of the publisher (Springer-Verlag, 175 Fifth Avenue, New York, New York 10010, USA), except for brief excerpts in connection with reviews or scholarly analysis. Use in connection with any form of information storage and retrieval, electronic adaptation, computer software, or by similar or dissimilar methodology now known or hereafter developed is forbidden.
The use of general descriptive names, trade names, trademarks, etc. in this publication, even if the former are not especially identified, is not to be taken as a sign that such names, as understood by the Trade Marks and Merchandise Marks Act, may accordingly be used freely by anyone.

ISBN 0-387-97159-9 Springer-Verlag New York Berlin Heidelberg
ISBN 3-540-97159-9 Springer-Verlag Berlin Heidelberg New York

Foreword

Global attention in scientific, industrial, and governmental communities to traces of toxic chemicals in foodstuffs and in both abiotic and biotic environments has justified the present triumvirate of specialized publications in this field: comprehensive reviews, rapidly published progress reports, and archival documentations. These three publications are integrated and scheduled to provide in international communication the coherency essential for nonduplicative and current progress in a field as dynamic and complex as environmental contamination and toxicology. Until now there has been no journal or other publication series reserved exclusively for the diversified literature on "toxic" chemicals in our foods, our feeds, our geographical surroundings, our domestic animals, our wildlife, and ourselves. Around the world immense efforts and many talents have been mobilized to technical and other evaluations of natures, locales, magnitudes, fates, and toxicology of the persisting residues of these chemicals loosed upon the world. Among the sequelae of this broad new emphasis has been an inescapable need for an articulated set of authoritative publications where one could expect to find the latest important world literature produced by this emerging area of science together with documentation of pertinent ancillary legislation.

The research director and the legislative or administrative adviser do not have the time even to scan the large number of technical publications that might contain articles important to current responsibility; these individuals need the background provided by detailed reviews plus an assured awareness of newly developing information, all with minimum time for literature searching. Similarly, the scientist assigned or attracted to a new problem has the requirements of gleaning all literature pertinent to his task, publishing quickly new developments or important new experimental details to inform others of findings that might alter their own efforts, and eventually publishing all his supporting data and conclusions for archival purposes.

The end result of this concern over these chores and responsibilities and with uniform, encompassing, and timely publication outlets in the field of environmental contamination and toxicology is the Springer-Verlag (Heidelberg and New York) triumvirate:

Reviews of Environmental Contamination and Toxicology (Vol. 1 in 1962 as *Residue Reviews* through Vol. 97 in 1986) for basically detailed review articles concerned with any aspects of chemical contaminants, including pesticides, in the total environment with their toxicological considerations and consequences.

Bulletin of Environmental Contamination and Toxicology (Vol. 1 in 1966) for rapid publication of short reports of significant advances and discoveries in the fields of air, soil, water, and food contamination and pollution as well as methodology and other disciplines concerned with the introduction, presence, and effects of toxicants in the total environment.

Archives of Environmental Contamination and Toxicology (Vol. 1 in 1973) for important complete articles emphasizing and describing original experimental or theoretical research work pertaining to the scientific aspects of chemical contaminants in the environment.

Manuscripts for *Reviews* and the *Archives* are in identical formats and are subject to review, by workers in the field, for adequacy and value; manuscripts for the *Bulletin* are also reviewed but are published by photo-offset to provide the latest results without delay. The individual editors of these three publications comprise the joint Coordinating Board of Editors with referral within the Board of manuscripts submitted to one publication but deemed by major emphasis or length more suitable for one of the others.

Coordinating Board of Editors

Preface

Despite attempts by the media to convince us our surroundings are under continual chemical assault and not faring well, there is abundant evidence that most chemicals are degraded or dissipated in our not-so-fragile environment. Yet, we must content with leaking underground fuel tanks, movement of nitrates and nitrites into our groundwater reservoirs, increasing air pollution in our large cities, and seemingly frequent contamination of our food and animal feeds with pesticides, industrial chemicals, and bacterial toxins.

Without continuing surveillance and intelligent controls, some of these chemicals could at times conceivably endanger the environment, wildlife, and the public health. Ensuring safety-in-use of the many chemicals involved in our highly industrialized culture is a dynamic challenge, for the old established materials are continually being displaced by newly developed molecules more acceptable to environmentalists, toxicologists, and federal and state regulatory agencies.

These matters are of genuine concern to increasing numbers of governmental agencies and legislative bodies around the world, for some of these chemicals have resulted in a few mishaps from improper use. Adequate safety-in-use evaluations of any of these chemicals persisting into our air, drinking water, and foodstuffs are not simple matters, and they incorporate the considered judgments of many individuals highly trained in a variety of complex biological, chemical, food technological, medical, pharmacological, and toxicological disciplines.

It is hoped that *Reviews of Environmental Contamination and Toxicology* will continue to serve as an integrating factor both in focusing attention upon those matters requiring further study and in collating for variously trained readers present knowledge in specific important areas involved with chemical contaminants in the total environment. This and previous volumes of "Reviews" illustrate these objectives. Because manuscripts are published in the order in which they are received in final form, it may seem that some important aspects of analytical chemistry, bioaccumulation, biochemistry, human and animal medicine, legislation, pharmacology, physiology, regulation, and toxicology are being neglected. To the contrary, these apparent omissions are recognized, and some pertinent manuscripts are in preparation. However, the field is so large and the interests in it are so varied that the editor and the Editorial Board earnestly solicit suggestions of topics and authors to help make this international book-series even more useful and informative.

Reviews of Environmental Contamination and Toxicology attempts to provide concise, critical reviews of timely advances, philosophy, and significant areas of accomplished or needed endeavor in the total field of foreign chemicals in any segment of the environment, as well as toxicological implications. These reviews are either general or specific, but properly they may lie in the domains of analytical chemistry and its methodology, biochemistry, human and animal medicine, legislation, pharmacology, physiology, regulation, and toxicology. Certain affairs in the realm of food technology concerned specifically with pesticide and other food-additive problems are also appropriate subject matter.

The justification for the preparation of any review for this book-series is that it deals with some aspect of the many real problems arising from the presence of any "foreign" chemicals in our surroundings. Thus, manuscripts may encompass those matters in any country. Added plant or animal pest-control chemicals or their metabolites that may persist into food and animal feeds are within this scope. The so-called food additives (substances deliberately added to foods for flavor, odor, appearance, and preservation, as well as those inadvertently added during manufacture, packing, distribution, and storage) are also considered suitable review material. In addition, chemical contaminant in any manner to air, water, soil, or plant or animal life are within this purview and these objectives.

Manuscripts are normally contributed by invitation but suggested topics are welcome. Preliminary communication with the editor is recommended before volunteered reviews are submitted in manuscript form.

College of Agriculture G.W.W.
University of Arizona
Tucson, Arizona

Table of Contents

Foreword v
Preface vii

Indoor Air Radon
   C. RICHARD COTHERN 1

Ecological Toxicology and Human Health Effects of Heptachlor
   E.A. FENDICK, E. MATHER-MIHAICH, K.A. HOUCK,
   M.B ST. CLAIR, J.B. FAUST, C.H. ROCKWELL,
   and M. OWENS 61

Subject Index 143

Indoor Air Radon

C. Richard Cothern*

Contents

I. Introduction 2
   A. General Properties 3
   B. Nuclear Properties 3
   C. Physical Properties 5
   D. Units of Radioactivity 9
II. Epidemiological Studies of the Health Effects of Radon 12
   A. General Comments 12
   B. US Uranium Miners 12
   C. Ontario (Canada) Uranium Miners 14
   D. Czechoslovakian Uranium Miners 15
   E. Saskatchewan Uranium Miners 17
   F. Swedish Iron Miners 18
   G. Newfoundland Fluorspar Miners 19
   H. Secondary Epidemiology Studies 20
III. General Comments on Epidemiology Studies 24
   A. Exposure 24
   B. Possible Interaction Between Smoking and Radon 25
   C. Other Variables 28
   D. Histopathology 30
   E. Other Health Endpoints 31
   F. Threshold? 32
   G. Future 33
IV. Dosimetry 33
   A. Anatomy of the Lung 34
   B. Properties of Alpha Particles 34
   C. Modeling of Deposition and Dose Equivalent 34
   D. Factors Involved in Dose Calculations 35
   E. Dose Equivalent Estimates 37
V. Animal Studies 37
VI. Risk Estimates 38
   A. Descriptive Models 38

---

*Executive Secretary, Science Advisory Board (A-101F), US Environmental Protection Agency, Washington, D.C., 20460. The thoughts and ideas in this review are those of the author and are not necessarily those of the US Environmental Protection Agency.

© 1990 Springer-Verlag New York Inc.
Reviews of Environmental Contamination and Toxicology, Vol. 111.

B. Population Risks	41
C. Discussion	45
VII. Risk Communication	47
Summary	49
References	50

I. Introduction

The presence of radon is a cause for concern, because it is radioactive; and inhalation of its decay products, or progeny, at sufficiently elevated levels is known to cause lung cancer.

Compared to other environmental contaminants, the information for radon is extensive, and for some needs it is perhaps a case of too much information. Because of the range of detail and the completeness of the data, there is often a problem in sorting out the different studies for situations where they differ or are in conflict. There are so many data, in fact, that different conclusions can be reached when different data are used.

The main source of indoor radon is soil, and it enters homes through cracks in basements, through holes around loose fitting pipes, and through holes in foundations. Radon also enters the home from drinking water (see Cothern 1987, Federal Register 1986).

In some instances the risk of developing lung cancer from exposure to radon can be equivalent to smoking several packs of cigarettes a day. The main risk of lung cancer, however, is from smoking cigarettes. It is estimated that more than 100,000 cigarette smokers die each year from lung cancer, and many more die from other smoking related diseases. In addition, it is estimated that 5000 lung cancer fatalities occur each year as a result of the effects of passive smoking. It is estimated here that between 4,000 and 30,000 fatal lung cancers occur each year due to exposure to radon in indoor air. The total of all lung cancer fatalities is presently about 130,000 each year in the United States.

The geological factors controlling radon occurrence can be grouped into three broad categories that bear upon the characteristics of the nearby geological formations: their radium (or uranium) content; their physical characteristics, and the degree of fracturing or faulting. These factors determine the amount of radon that will be produced and enter into the soil gas and how easily the gas will move through the soil. The amount of radon in soil gas and the permeability of the surficial material are probably the most significant natural factors affecting indoor radon concentrations.

This review is not exhaustive and does not necessarily cover every detail concerning radon and its progeny but it is intended to give an overview of environmental radon and the risk it brings to humans. The main emphases of the current review are the epidemiological studies and an analysis of the related data. For a more complete, comprehensive discussion the reader is referred to a supplementary textbook entitled "Environmental Radon" (Cothern and Smith 1987). For other lists of references see the recent bibliography (USEPA 1988b). Some recent reviews have appeared that cover some of the same material (NIOSH 1987, IARC 1988).

Some background is provided here concerning general nuclear decay and physical properties, and units of measurement. The review is organized around the data and analysis needed to determine the individual risk rate and the population risk (total number of lung cancer fatalities due to radon) caused by radon in the US.

The next section of this introductory chapter is a discussion of some basic nuclear physics and general properties of radon followed by a discussion of the units used to describe radon concentrations. Depending on your background you may wish to skip part of this section.

A. General Properties

Radon is a naturally occurring, odorless, colorless, almost chemically inert, radioactive gas. Compared to the other noble gases, radon is the heaviest of these and has the highest melting point. It is soluble in cold water, and its solubility decreases with increasing temperature. Although radon is an inert gas it forms some compounds such as clathrates and complex fluorides under special conditions. Efforts have been unsuccessful to form oxides and other halides with radon.

B. Nuclear Properties

An atom consists of a heavy concentration of mass at the center (the nucleus) surrounded by shells of electrons in different orbits. The primary constituents of the nucleus are neutrons and protons. The neutrons have no charge while the protons have a positive charge. The orbital electrons have a negative charge and are equal in number to the protons, making the atom neutral in overall charge.

The characteristic of the noble gases can be understood using the idea of electron orbits. The noble gases have completely filled outer electron orbits. If the first orbit is filled, the atom is helium (He). When the orbits are each completely filled, the atom has greater stability or is less reactive, hence the inert gases. If the first and second orbitals are filled, the element is neon. This sequence continues through Argon (Ar), Krypton (Kr), Xenon (Xe), and Radon (Rn).

The number of protons in the nucleus determines the element and its atomic number. A given element can have more than one particular number of neutrons. Variation in the number of neutrons does not change the chemical properties (the element is the same), but it produces considerable change in the stability of the element to radioactive decay. Atoms with the same number of protons but different numbers of neutrons are called isotopes. For example, if an atom has 86 protons, it is radon. There are three well known isotopes of radon containing 133, 134, and 136 neutrons, respectively, and a total of 24 other isotopes. The atomic mass number is the total number of protons and neutrons in the nucleus and this sum is usually used to label isotopes. The three isotopes of radon have atomic masses of 86 + 133 = 219, 86 + 134 = 220, and 86 + 136 = 222. Symbolically there can be written as: ^{219}Rn, ^{220}Rn, and ^{222}Rn, or Rn-219, Rn-220, and Rn-222.

It is a general rule of nature that a system will try to attain the lowest energy state or the most stable situation possible; e.g., water runs downhill, or unlike charges attract each other causing an electron to "fall" into the orbit closest to the nucleus. In this same sense, if a nucleus can move to a lower energy state by emitting radiation, it will. Such a nucleus is radioactive compared to other nuclei which may be stable, and unable to lose energy by emitting radiation.

In general one might expect the nucleus to be able to emit all different kinds and combinations of radiations. However, because of this trend to stability and the nature of the nuclear forces, the most stable radiations to be ejected are:

Emitted particles	Process	Radiation type
Helium nucleus (two protons plus two neutrons)	Alpha decay	Alpha particles
Electron	Beta decay	Beta particle
A kind of high energy X-ray	Gamma decay	Gamma ray

An alpha particle, the heaviest nuclear radiation, consists of two protons and two neutrons (a proton or neutron is about 2,000 times as massive as an electron). A beta particle is an electron emitted from the nucleus as a result of neutron decay. An electron can be "created" and ejected from a nucleus by a neutron decaying into a proton (which remains in the nucleus) and an electron (which is ejected as a beta particle). As a result of this process the nucleus has one more proton and thus has become the atom of a different element with atomic number one greater than the parent atom. A gamma ray is a form of electromagnetic radiation. Other forms of electromagnetic radiation are light, radio waves, infrared radiation, ultraviolet radiation and X-rays.

The processes of alpha and beta radioactive decay lead to a different element, while gamma decay does not. The decaying isotope is known as the parent. The resulting isotope (if a different element) is the progeny (historically referred to as a daughter). For example, ^{222}Rn decays by emitting an alpha particle to the progeny ^{218}Po. This reaction is written:

$$^{222}\text{Rn} \rightarrow {}^{218}\text{Po} + {}^{4}\text{He}$$

Beta decay causes the atomic number to increase by one. The process can be described as a neutron in the nucleus being converted to an electron and proton with the ejection of an electron. An example of beta decay is the decaying of ^{228}Ra to ^{228}Ac. This is written:

$$^{228}\text{Ra (atomic number 88)} \rightarrow {}^{228}\text{Ac (atomic number 89)} + \text{beta particle}$$

Alpha, beta, and gamma radiations have many different energies and masses and thus produce different effects as they interact with matter. Each of these radiations is capable of knocking an electron from its orbit around the nucleus and away from the atom, a process called ionization. If an electron is moved to an orbit further from the nucleus the atom is said to be excited. The atom will then decay by the electron returning to the inner orbit and emitting radiation. This process is used in most methods for detecting and measuring radiation.

Not all atoms are equally stable, and different isotopes characteristically decay at different rates. The concept of half-life is used to describe these differences quantitatively. The half-life of an isotope is the time required for one-half the atoms present to decay. Half-lives can range from billions of years (the half-life of ^{238}U is 4.5×10^9 years) to millionths of a second (the half-life of ^{214}Po is 164×10^{-6} sec) and even less.

Another way to describe the differences between the nuclear radiations is their ability to penetrate matter. In general most alpha particles can be stopped by a piece of paper, or more precisely, by the mucociliary blanket of the lung, a range of 30–50 μm in thickness. That the alpha particle can be stopped in such short distances, indicates that it deposits high energy in a small distance, causing more damage per unit volume than other radiations.

Many isotopes exist naturally, such as the ^{40}K in humans, the ^{14}C produced by cosmic rays used to date old manuscripts, and the naturally radioactive series shown in Figs. 1, 2, and 3. There are three naturally occurring radioactive series: the uranium, thorium, and actinium series. These series involve a sequence of alpha, beta, and gamma decays with heavy nuclei. They start respectively with ^{238}U, ^{232}Th, and ^{235}U, and all end with a different stable isotope of lead (Pb). In the middle of each series a different isotope of the gas radon (Rn) is formed. The radon isotope in the uranium series is called *radon*, that in the thorium series is called *thoron* and that in the actinium series is called *actinon*. The implication of a gas being formed is important to human health, since gases have considerable freedom to move.

C. Physical Properties

When radon is inhaled with indoor air, it remains in the lungs for only a short time, and then most of it is exhaled. When radon first decays in indoor air, it produces another atom, which has an electrostatic charge for a very short time, since the exiting alpha particle removes more than two electrons. The resulting ion, ^{218}Po, is very reactive and is electrostatically attracted to tiny particulates (e.g., small aerosols, small Aitken particles and other Bunsen particles with dimensions less than 0.7 μm) which are inhaled and may be small enough to move past the body's natural defenses and be deposited in the lung. The progeny of radon then decay sequentially, releasing damaging nuclear (alpha, beta, or gamma) radiations. Thus it is the progeny of radon that actually cause the damage

THE URANIUM SERIES

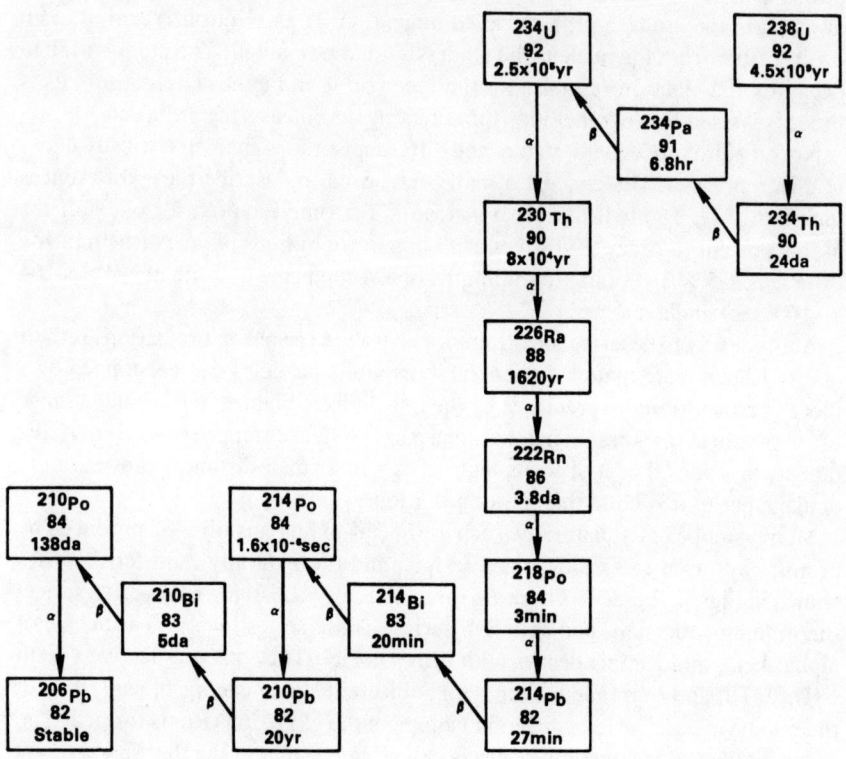

Fig. 1. Schematic representation of the naturally occurring uranium radioactive series. The isotope of radon, radon-222 or radon, is shown as the immediate decay product of radium-226. Times are half-lives, superscripts on chemical symbols are the atomic masses, and subscripts are atomic numbers. Alpha indicates an alpha decay and beta indicates a beta decay.

THE THORIUM SERIES

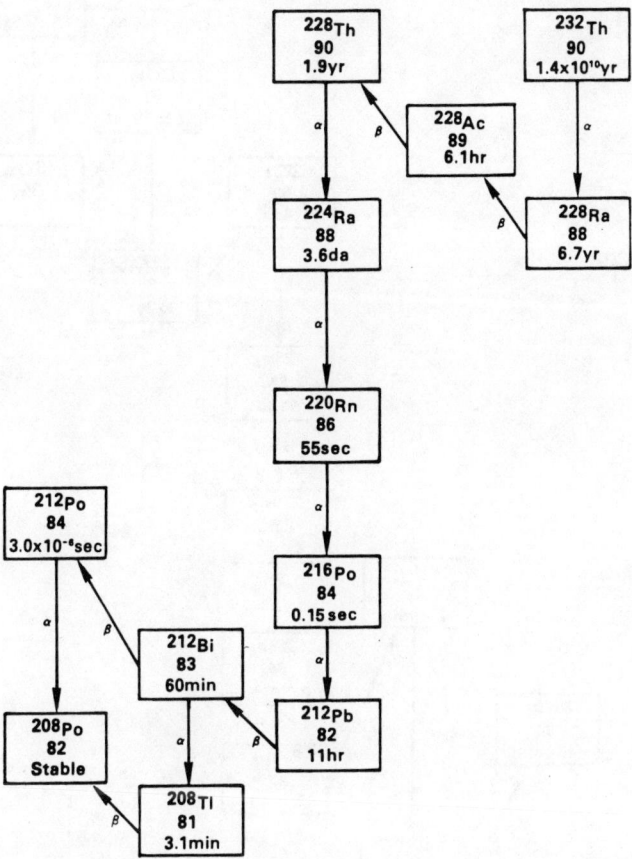

Fig. 2. Schematic representation of the naturally occurring thorium radioactive series. The isotope of radon, radon-220 or thoron, is shown as the immediate decay product of radium-224. Times are half-lives, superscripts on chemical symbols are the atomic masses, and subscripts are atomic numbers. Alpha indicates an alpha decay and beta indicates a beta decay.

THE ACTINIUM SERIES

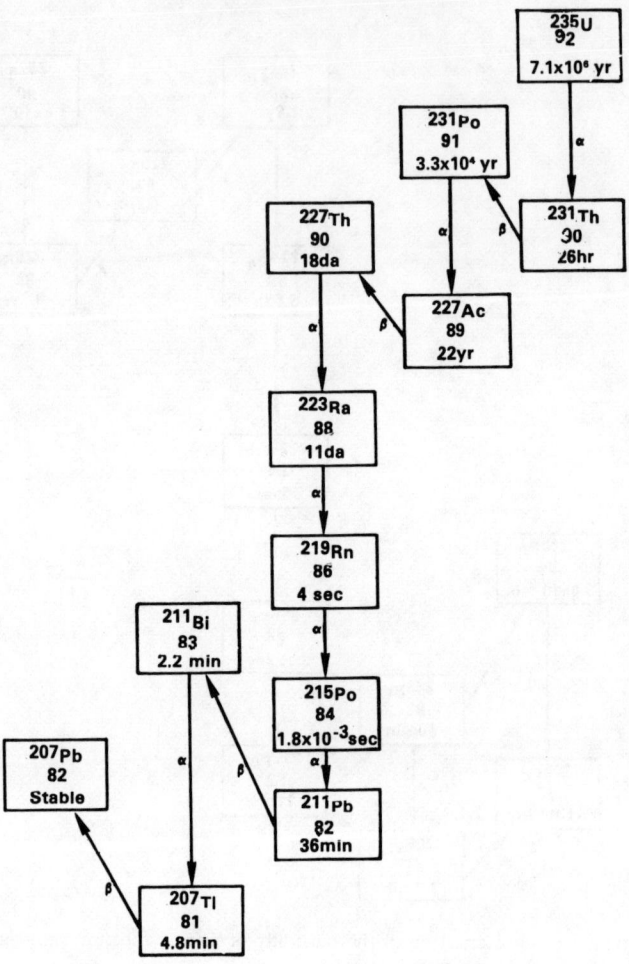

Fig. 3. Schematic representation of the naturally occurring actinium radioactive series. The isotope of radon, radon-219 or actinon, is shown as the immediate decay product of radium-223. Times are half-lives, superscripts on chemical symbols are the atomic masses, and subscripts are atomic numbers. Alpha indicates an alpha decay and beta indicates a beta decay.

to bronchial epithelium and can induce lung cancer, because only the progeny remain in the lungs for a time adequate to result in significant decay.

The progeny that are likely adsorbed on particulates are known as the attached fraction and those isotopes of the progeny of radon that remain atomic or molecular in nature, or are surrounded by water or other molecules, are referred to as the unattached fraction. In time the activity of the progeny will grow, and for times long compared with the half-lives involved, the activity of all progeny will be equal. However, in general, sufficient time has not elapsed for equilibrium to be fully realized, and any estimates of the concentration of the different progeny must take this into account.

D. Units of Radioactivity

Generally units such as mg/L, µg/L, or parts per million (ppm) are used to describe the concentration of pollutants and hazardous substances. However, certain unique properties of radioactive substances preclude the use of these units and require different units to directly compare the health effects of different radionuclides.

Three important units are needed to describe radioactivity and answer the following questions:

- How many radiations/sec or decays/sec or disintegrations/sec are emitted?
- How much punch does the tissue receive or how much energy is imparted to matter (the dose)?
- How much biological damage is produced by the radiation?

Historically units were developed to describe these properties and called activity, dose, and dose equivalent. They are curies, rads, and rems, respectively. Recently radiological units have been added to the International System (SI). The units corresponding to the above are *becquerels*, *grays*, and *sieverts*, respectively. In the descriptions below, and in this review, both sets of units are used, although it should be recognized that there is a move in the health physics community toward the SI units.

For radioactivity, the number of particles (alpha, beta or gamma) emitted causes the damage and not the mass of the radionuclides. Thus it is essential to have a unit that describes the activity or number of particles emitted. The activity is inversely related to the half-life, and thus longer half-lives mean lower activity. Historically, by definition, 1 g of radium (^{226}Ra) is said to have 1 curie (1 Ci) of activity. By comparison 1 g of ^{238}U has an activity of 0.36 µCi. The SI unit of activity is the becquerel (Bq) which is defined as one disintegration/sec. For conversion purposes 1 Bq = 27 picocuries (10^{-12}) (pCi) (see Table 1 for others).

The historical unit of dose, or radiation absorbed, is the rad. One rad deposits 100 ergs (the unit of energy in the cm/g/sec metric system) in 1 g of matter. To get a better perspective on the size of an erg, 10 million ergs/sec is one watt. The

Table 1. Some Conversion Factors for Radon and Radon Progeny

1 WL = 100 pCi/L (at equilibrium)
= 200 pCi/L (approximate indoor concentration)
1 Bq = 27 pCi
1 pCi/L = 37 Bq/m^3
1 Jh/m^3 = 285 WLM
1 WLM = 0.0035 Jh/m^3
1 WL = 2.08 × 10^{-5} J/m^3
= 51.6 WLM/year
1 WLM/year = 4.03 × 10^{-7} J/m^3

SI unit for dose (defined in the m/kg/sec metric system) is the gray (Gy,) and it can be thought of as depositing 1 joule of energy per kilogram (J/kg) of matter. To convert between the historical and SI systems the relationship 1 Gy = 100 rad is used.

Because of the importance of a particle's mass and charge, 1 Gy, or 1 rad of alpha particles creates more damage than 1 Gy or 1 rad of gamma rays. To compensate for this difference in effect a different unit was invented, the sievert (SI) or rem (historical). This is known as the dose equivalent. The dose is measured in Gy or rads and the dose equivalent is measured in sieverts or rems. Frequently for convenience, however, the sievert (Sv) or rem is also used for dose. The dose equivalent is a measure of harm and is not generally an exact measurement; it is a useful accounting unit. The dose and dose equivalent are related by a quality factor as follows:

$$Sv = Q \times Gy$$

where Q is the quality factor assigned the following value:

$Q = 1$ for beta particles, gamma ray and X-rays
    = 10   for neutrons from spontaneous fission and protons
    = 20   for alpha particles and fission fragments

The SI unit for dose equivalent is Sv, which is in the m/kg/sec metric system a J/kg. To convert between the SI and historical systems use this relationship: 1 Sv = 100 rem.

The effect of ionizing radiation on different organs is not the same. In order to include this variable, a system of measure has been developed that weights the dose equivalent by its potential or harmfulness to a particular organ. The resulting quantity is called the effective dose equivalent. This quantity was developed by the International Commission for Radiation Protection (ICRP) and when their weighting factors are used the resulting effective dose equivalent values are directly applicable for occupational exposures.

The average annual effective dose equivalent in the US population is shown in Table 2, which gives some idea of the magnitude of the unit. The average human

Table 2. Exposure of the US Population to Ionizing Radiation from Various Sources (From NCRP 1987)

Source	Average Annual Effective Dose Equivalent in the US Population (mSv)
Natural sources	
Cosmic	0.27
Cosmogenic	0.01
Terrestrial	0.28
Inhaled radon progeny	2.00
In the body	0.39
Occupational	0.009
Nuclear fuel cycle	0.0005
Consumer products	0.05 − 0.13
Miscellaneous environmental sources	0.0006
Medical	
Diagnostic x-rays	0.39
Nuclear medicine	0.14
Rounded Total	3.6

receives from cosmic rays (high energy protons from outside the earth), and natural radiation other than radon, an effective dose equivalent of about 1 mSv/yr. This can vary depending on where one lives and the kinds of structures in which one lives and works. The higher the altitude, the less protection from the earth's atmosphere. Thus residents in Leadville, Colorado, receive from cosmic rays 1.1 mSv/yr, while those at sea level receive about 0.2 mSv/yr. Flying coast-to-coast can add as much as 0.05 mSv per flight.

Until recently the contribution of radon to the natural background was not listed in most compilations. Thus no mention is found in earlier estimates of its contribution. It is now believed to be in the order of 2 mSv/yr or one-half the total average natural background radiation exposure (NCRP 1984).

A selected population is the US is subjected to diagnostic X-rays and that process contributes an additional 0.50 mSv/yr. A similar group will receive additional exposure to ionizing radiation, as in cancer treatment. Those who receive radioiodine treatment for a thyroid condition can receive as much as 20 mSv. Color TV can lead to exposures as high as 0.01 mSv/yr. The effluents from nuclear power plants may contribute a small fraction of a mSv/yr. Exposure to dental X-rays and occupational exposure to some professions contribute additional doses. From the sum of these, the population in the US is exposed to an approximate dose equivalent of 3.6 mSv/yr (or roughly 360 mrem/yr) as shown in Table 2.

Since it is the progeny of radon that cause the lung damage, the concentrations of progeny in the air should be measured rather than the concentrations of radon. There are four major progeny isotopes, and measuring the concentration of each is somewhat complex and difficult. A compromise is reached, however, by inventing a unit that is both descriptive of the combined progeny concentration and easily measured. This unit, the working level (WL), has been used historically for measuring radon progeny in mines.

One WL is defined formally as the combination of radon progeny in 1 L of air that results in the emission of 1.3×10^5 MeV of alpha particle energy. If all the progeny are in equilibrium with the parent radon, then one WL is numerically equal to a radon concentration of 100 pCi/L of air. Equilibrium, however, is seldom achieved. For calculation purposes it is often assumed that, on the average, approximately 50% equilibrium is achieved in indoor air. In that situation, 1 WL would be equal to 200 pCi/L of indoor air. Therefore the WL unit describes the progeny of radon, whereas the pCi/L unit describes the concentration of the parent radon in air. The degree of exposure occurring over time is described in terms of exposure to one WL for an occupational mon of 170 hr, which is 1 WLM. (See Table 1 for conversions between historical and SI.)

II. Epidemiological Studies of the Health Effects of Radon

This section surveys the epidemiological studies of the health effects of radon for cohorts involving miners and other groups. General characteristics are detailed first, and then comments are added for the major studies that reflect judgments from both the scientific literature and from interviews with their principle investigators. Generic judgments from those collecting and analyzing the data can be found in section III.

A. General Comments

In the past 10–20 yr substantial epidemiological information on lung cancer risk among several populations has become available. In general, those studies that provide dose–response data predict essentially the same risk levels. Some researchers would argue that the strongest of these studies are for US and Canadian uranium miners and Czechoslovakian miners. Studies of Saskatchewan, Newfoundland, and Swedish miners are thought by some to be weaker but also provide quantitative dose–response information. The Colorado, Czechoslovakian, and Ontario studies appear to be internally consistent, while the other three have internal inconsistencies. There are several other studies which are more qualitative, and are discussed.

B. US Uranium Miners

1. Background. The epidemiological studies of US metal miners in Colorado began in the 1960s using records of 75,000 miners, among whom 1,759 were

Table 3. Summary of Standard Mortality Ratios and Attributable Risks for Epidemiology Studies of Radon Exposure

Study	SMR	Excess Lung Cancer Cases per 10^6 Person-Years WLM	
Swedish iron ore	288	13.0–43.2	50–65 yr old
		27.2–90.2	> 65 yr old
US metal	292		
UK tin	211		
UK iron	174		
France uranium	191		
Chinese tin	1451		
Norway niobium		50	
US uranium	482	9	
Czechoslovakian	500	11.5–44.2	
	(150–900)		
Swedish zinc		30	
Newfoundland	426	5.5–6.0	
Ontario uranium	181	17	
Swedish iron	390	21.0 for smokers	
		19.0 for non-smokers	
Saskatchewan	190	20.8	
Northwest Canada	230	3.1	
Jacobi (1981)		5–15	
Thomas et al. (1985)		3.4–17.8	

chosen who had > 15 yr mining experience between 1937 and 1948 (Wagoner 1963). The first report on the uranium miner studies was in 1962 and involved 3,306 miners exposed as of 1957. These studies have been periodically updated involving the medical examination of 3,362 white and 780 non-white miners who worked at least 1 mon as of January 1, 1964, and were followed through December 31, 1977 (Lundin et al. 1971; Waxweiler et al. 1981; NIOSH review 1987). The exposure for this group was skewed with only a few receiving very high doses. In the 1977 update, 185 lung cancer deaths were documented among 950 who had died, compared to 38.4 expected lung cancer deaths, giving a standard mortality ratio (SMR) of 482 ($p < 0.05$) involving 65,556 person years at risk. See Table 3 for a comparison of the SMRs of many of the epidemiology studies. The analysis was conducted using the formula for attributable risk, and this determined that 80% of the lung cancer deaths were attributable to uranium mining. Statistically significant excess cancer rates were found in all categories above 120 WLM.

2. Comments. This study is the best and most extensive for smoking data, although no data are available after 1969. About 70% of the miners were smokers, which is about twice the value for the general public, although there were fewer heavy smokers consuming more than 1.5 packs/d (Cote and Townsend 1981). The incidence of lung cancer in the nonsmokers in this cohort, as the analysis is extended in time, appears to be catching up with that of the smokers. Thus it is concluded that smoking only causes the lung cancer to occur at an earlier age.

The study used death certificates and in many cases, especially for early exposures, the exposure information was estimated from data for nearby mines (Lundin 1969). There was considerable job turnover, with the majority of miners working < 10 yr and several miners working in uranium and other mines prior to the study. Although it was estimated that there was only a 16% contribution to exposure from prior work, the bias was toward overestimating exposure levels. This was because a disproportionate number of mines were from areas with high radon progeny levels and sampling was done more in mines with high levels than in those with low levels. About 10% of the early working level exposure numbers are based on actual measurements (Lundin 1971). In general, this study gives lower estimates than the other epidemiological studies (NAS 1980, NCRP 1984). Another complication in the exposure estimates is that due to improved mining practices including ventilation. The average annual potential alpha energy concentration initially came down from approximately 15 WL in the 1940s, to a range of 6 to 10 WL in the 1950s, and then after 1970, to something below 0.5 WL today. However, the measurements over the years have been carefully conducted and are among the most reliable. About one-fourth of the exposure values are now based on measured values, about one-fourth are extrapolated from similar mines, and the rest are "guesstimates." If anything, the exposure values are overestimated, because they were often used when the mine was involved in an enforcement action. In an early study of the quality of the exposure measurements taken by mine operators, the actual measurements produced results that were four times greater than those recorded by the mine operators, while more recent data show this difference reduced to a factor of two. More improvement does not seem possible. In the early years the exposure values were likely higher, since it was suspected that the mine owners did corrective action while the inspector waited to measure radon levels.

The consequences of exposures to other contaminants are not yet quantified, and there is possible risk from aromatic hydrocarbons, alkylating agents, metal dusts, soot from diesel engines, zinc and lead from blasting primers, sulfur from powder fuses, and welding fumes (Cote and Townsend 1981; Axelson and Sundell 1978; Steinhausler 1988).

C. Ontario (Canada) Uranium Miners

1. Background. The cohort of uranium miners in Ontario, Canada, included 15,984 who worked at least 1 month and entered the mines between January 1, 1955, and December 31, 1977. Most of the miners in this group worked only for very short periods. In general the exposure levels were very low and in the range of 40 to 90 WLM (Muller et al. 1981, 1985). To date, 119 lung cancer deaths have been recorded with 66 expected, giving a SMR of 181 ($p < 0.001$). The SMR = 118 for all cases and = 223 for violent mining deaths. This cohort is very young at this time.

An increased risk of stomach cancer was seen among Ontario gold miners but not among any other Ontario miners (Muller 1985). The exposure levels to arsenic, nickel and cobalt are expected to be quite low (Chambers 1985).

2. Comments. Among the problems with this study is that approximately 66% of the miners had previous experience and thus other exposures. The exposure levels were based on area monitoring, and it may well underestimate the actual individual levels (Muller et al. 1983). Since the follow-up period was truncated at 18 yr, and the median age of the cohort was only 39 yr; the known long latency period introduces an uncertainty. It is possible that the thoron levels contributed significantly to the exposure, and that thoron could have been mistaken for radon; because the measurement method had not properly modified a Kusnetz method used to determine the thoron concentration (Cote and Townsend 1981). The complication is that there are no epidemiological studies for thoron.

The exposure levels are known to have varied widely and few measurements were actually made. Exposure was often calculated from "best estimates" rather than a single value for each miner. After 1967, individual measurements were initiated. These estimates may have been underestimated (Thomas et al. 1985).

This group had a higher mobility than the others such as the Czech or US miners. There were no data presented concerning levels of arsenic, asbestos, blasting fumes, radioactive ore dust, or diesel exhaust fumes. Also no data were available on histopathology of the lung cancer tissue and there was some confusion regarding the location of the tumors (trachea, bronchal, lung, larynx, etc.) between the miner and control population.

D. Czechoslovakian Uranium Miners

1. Background. The dose–response studies of Czechoslovakian uranium miners, although these mines are likely the oldest known, began only in the 1970s with data analyzed to 1973 at first (Svec et al. 1976). These data include perhaps the best records for radon exposure and may have measured equilibrium values since 1966 (Svec et al. 1976). By 1978 the group had shown that there were statistically significant levels of lung cancer in all categories greater than 100 WLM (Kunz et al. 1978). The group has ruled out confounding factors such as silica, arsenic, asbestos, chromium, nickel, and cobalt due to their variations in levels between different mines (Svec et al. 1984).

In this study small cell undifferentiated and epidermoid type cells were seen in the histology studies. Basal cell carcinomas of the skin developed in the miners.

Table 4. Attributable Risk for Czechoslovakian Uranium Miners

Group	Mean WLM	Population yr at risk	Observed	Expected	Attributable risk 10^6/WLM/yr
S (2 Groups)	229.0	45501	135	14.8	11.5
	224.0	39557	349	82.7	31.0
N (2 Groups)	5.2	37127	0	0.8	–
	5.6	6873	4	2.3	44.2
K	40.1	13204	9	2.8	11.7
L	23.4	8731	22	14.2	38.2

These cancers may be due to exposure to arsenic or due to external doses of radiation from radon progeny with cumulative doses exceeding 1000 rem. The attributable risk level for these carcinomas is approximately 10^{-4} persons/yr/Sv (Svec et al. 1988).

The data showed that about 70% of the miners smoked cigarettes, but this is comparable to the general male population.

In 1988 the group reported that they had completed four epidemiology studies involving six miner groups of which some had been partially studied before and some were new. These studies involved uranium, iron and shale clay miners. These studies showed a significant excess of lung cancer at exposure categories below 50 WLM and the first excess lung cancer rate appeared in the sixth year following exposure as seen in Table 4.

In a later analysis, the modified life table technique was used (Svec et al. 1988). The exposures in the newer studies used personal dosimetrical cards with direct measurements of radon progeny. The dosimetry of the older studies was re-evaluated. Although an earlier study of this group had shown no influence due to cigarette smoke, this newer analysis gave the rate ratio between smokers and non-smokers as 1.6. The observed to expected death rate ratio for malignant neoplasms other than lung cancer was < 1.0.

The analysis of the above data showed a statistically significant excess in observed rates of lung cancer in the exposure category with a mean of 79 WLM. For miners with the onset of exposure after 30 yr of age an excess was found in the exposure category with a mean of 39 WLM. Also the shape of the dose–response curve was different, being concave downward for the age group that started exposure after 30 yr of age (Svec et al. 1988). They report a decrease of risk coefficient with decreasing age at the start of mining and at high exposures with increasing exposure rate. Also the dose–response curves appeared to be different for epidermoid and small cell types of lung cancer.

2. Comments. This study is the nearest to the life span of the miners. It is a long-term study with sufficient number of deaths for statistical accuracy, however, not all the data are given in the publications to be able to fully evaluate its validity. Perhaps the biggest drawback to this study is the lack of full data on smoking. It is believed that some prisoners and slave labor have been used in these mines, and the length of the working day may have actually been longer than reported. Also the follow-up of such people is likely to be incomplete. This is reflected in the change in number of miners between 1971 and 1988.

There is some uncertainty introduced by the fact that the exposure measurements were of radon rather than radon progeny. There were insufficient data on equilibrium given (e.g., in Svec 1976) to be able to estimate radon progeny levels accurately. However, it has been estimated that this difference introduces an uncertainty of no more than 20%.

The cohort definition is enough different from other main studies to make direct comparison difficult. For instance, a miner had to work for 4 years to be included in the cohort. It was not clear if the time of entry into the cohort or time of first exposure was used and thus it is not possible to use a life-table analysis. Only 2% of the uranium miners had other hard rock mining experience. Some other researchers in the area have been suspicious of the small statistical variation in the data (Steinhausler 1988). The population of this group has been somewhat mobile and the subsequent changes of government have forced many to leave the country.

In comparison to controls, the Czechoslovakian uranium miners show a deficit of epidermoid tumors, no difference for adenocarcinomas and an excess of small cell anaplastic tumors (Horacek et al. 1977). Others have questioned these results, because of a lack of important histological data (Steinhausler 1988).

E. Uranium Miners in Saskatchewan, Canada

1. Background. The Eldorado Beaverlodge uranium mine in Saskatchewan, Canada, employed 8,487 mine workers between 1948 and 1980. In this cohort there were 65 lung cancer deaths compared to 34.24 expected (Howe et al. 1986). Exposures prior to 1966 were estimated, and after that time radon-progeny measurements are available. Silica exposures were very low, and diesel machinery was never used underground. The attributable risk was found to be $20.8/10^6$ person-years WLM, and the relative risk was found to be 3.28% per WLM. In this retrospective study the radon progeny concentration has been measured several times per month since 1967, and exposures were estimated from records of hours spent at each location. The equilibrium factor was determined by direct measurement. A significant risk was recorded below 50 cumulative WLM.

The Eldorado Port Radium uranium mine study involved a cohort study of 2,103 workers employed between 1942 and 1960 (Howe et al. 1987). A total of 57 lung cancer deaths were observed, yielding a relative risk coefficient of 0.27/WLM and an attributable risk coefficient of $3.10/WLM/10^6$ person-years.

2. Comments. These dose–response data from these Saskatchewan mines support a linear relationship, the death data were from Statistics Canada and there were no histopathology or smoking data, and no information on other confounding factors. These studies were for only a brief period of exposure, although the follow-up appears to be good and shows that those > 50 yr of age were at greater risk. The exposure estimates could be underestimated, because median values and not mean values were used. The miners are still very young and they need to be followed for several more years. Mortality data and not incidence data were used in these studies. This approach always overestimates exposure, since they are accumulating dose after the incidence of the lung cancer, sometimes called the "wasted radiation" concept. The exposure values are viewed by many with suspicion as they may be incomplete.

The data used were from the Canadian National Mortality Data Base, and exposures are likely underestimates as there is no information before 1940. Only natural ventilation was used before 1947, and equilibrium factors for that period were estimated. No smoking data were available, and no information was available on other radon exposures. Because of the many problems with the estimates of exposure, these data are more suspect than the others. It is suggested that the difference between these miners and those at Beaverlodge is a true biologic difference, that exposure at a higher rate gives a lower attributable and relative risk.

F. Swedish Iron Miners

1. Background. This cohort of Swedish iron miners involves 1,415 men who were born between 1880 and 1919, were alive in 1930, and worked more than one calendar year between 1897 and 1976 (Radford and Renard 1984). Between January 1, 1951, and December 31, 1976, there were 50 lung cancer deaths in this cohort while 14.6 were expected giving a SMR of 342 ($p < 0.01$). When adjusted for smoking the expected deaths decreased to 12.8 and the SMR increased to 390 with p likely < 0.05. There were statistically significant lung cancer death rates for exposure levels > 80 WLM ($p < 0.05$). In Sweden roughly 50% of deaths are followed by autopsy, and 70% of the lung cancers in this cohort were confirmed by autopsy or thoractomy. The attributable risk for smokers was 21.8 excess cases of lung cancer/10^6 person-years WLM and for non-smokers was 19.0 excess cases of lung cancer/10^5 person-years WLM. No significant excess of cases was observed until 20 or more years after the first exposure. It was concluded that the effect of smoking was additive; that latency was independent of smoking category and that lung cancer incidence is independent of age of first exposure.

2. Comments. This was a small study of an older group with ages in the 70s and over 70% dead. Some researchers are suspicious of this study, since it is consistently on the high side. The smoking data appear to be different from those of many other studies. There are more miners at lower doses, but it appears that too much extrapolation was involved in estimating the exposure. Over 90% of the exposure values were extrapolated, and many were used over a long period, when the ventilation conditions were changing.

The exposure records for this cohort were limited and the mean exposure was approximately 5 WLM/yr. Values were estimated based on ventilation assumptions, crude calculations, and tenuous connections between the exposure and cohort group. They extrapolated the exposure estimates from 1959 back to earlier dates. In some cases there is an unclear definition of the cohort; for example, it is not clear how the latency period was adjusted for persons at risk < 10 yr although this may be a minor effect (NIOSH 1987). These data are unusual in that all of the groups exposed at levels from 34 WLM to 220 WLM showed the same excess mortality; therefore, there is no dose–response in this range. This is possibly an example of a stratification effect.

The histology types were almost equally epidermoid and undifferentiated, and there was no evidence of an effect of smoking on the histologic findings.

Other problems with this study include the small numbers involved and the additional potential contribution to exposure from the indoor air radon levels in homes. Also the water from underground springs is in the range of 2 to 20 nanoCuries (nCi)/L and thus could be a major source of radon.

Diesel exhaust is not a likely confounder, since 70% of the lung cancer cases had left underground work or died before the introduction of diesel equipment in the 1940s. Respiratory carcinogens such as arsenic, chromium, nickel and asbestos were essentially absent.

G. Newfoundland Fluorspar Miners

1. Background. The mining of calcium fluoride or fluorspar in Newfoundland involved 2,120 miners, millers and surface workers who were employed between 1933 and 1978. Of this total cohort 56% were underground miners. In this mine a significant contribution to the radon levels came from contaminated ground water (Morrison et al. 1985). By 1977 there were 98 lung cancer deaths of which 89 were from the underground miners and 9 were from the surface workers. Also the miners were heavy smokers, with 86% smoking cigarettes (Morrison et al. 1981). By 1981, it had been observed that there were 104 lung cancer deaths where 24.4 were expected, giving a SMR of 426 (p estimated to be < 0.05). The attributable risk for this group is in the range of 5.5 to $6.0/10^6$ person-years WLM depending on the smoking status.

2. Comments. The exposure levels in this mine were unusually high (Morrison et al. 1985). There were infrequent and limited exposure measurements made before 1968, and they suggest that the average level was of the order of 0.5 WL (Morrison et al. 1981). Exposure levels were extrapolated back from 1959 and many of the estimates were purely speculation. The study failed to trace the whereabouts of a large number of workers. This study suffers from the confounding factors of including surface workers, although this may be a small effect, since only a small percentage of the total group were surface miners. The study may exaggerate the overall risk due to a lack of smoking histories. There could have been large differences in exposure levels within an individual mine and between mines. Natural ventilation was used before 1960, and the incidence data were collected from death certificates. Because of these difficulties, Steinhausler (1988) concludes that this study cannot be used for determining a risk assessment. From limited histology it was determined that 90% of the lung cancers were squamous cell carcinomas and 7% were small cell carcinomas.

H. Secondary Epidemiology Studies

These studies show a relationship between radon exposure levels and lung cancer, but for various reasons they are not usable for developing quantitative dose–response relationships. Some are in progress and will hopefully yield dose–response information.

1. Coal and Tin Miners in the UK. There are 185,200 underground coal miners in the UK exposed to typical radon levels of approximately 2 pCi/L (O'Riordan 1981).

Among the 1,333 tin miners in a cohort from Cornwall, UK, there were 28 lung cancer deaths compared to 13.27 expected, yielding a SMR = 211 with p estimated to be < 0.05 (Fox 1981; NIOSH 1987). There was no smoking information given and it was not clear if the miners were surface or underground miners. The SMR for deaths from respiratory diseases and silicotuberculosis was 473. In 1981 only 4% of non-coal miners in the UK were exposed to levels exceeding 4 WLM/yr (O'Riordan 1981).

2. Iron Ore Miners in Grangesberg, Sweden. Iron ore has been mined here since medieval times. It was started underground in 1910, and the ventilation was natural until after 1945. The cohort for the iron ore miners in Grangesberg, Sweden, was quite small with lung cancers being divided into two groups, (50–65 years old, and > 65 years). The cumulative exposure was 148.5 and 167.2 WLM, respectively. The risk estimates were 13.0–43.2 and 27.2–90.5 cases/10^6 person-years WLM. The overall risk estimate was 30–40 excess cases of lung cancer/10^6 person-years WLM. They calculated a SMR of 16.2 with $p < 0.0001$ (Edling and Axelson 1983). Parish and burial registers were used from the mining community of approximately 3,000 using only men who were over 50 years old. There is a large uncertainty in the exposure data. It was noted that there is a tendency for a shorter induction period for smokers, and an additive effect is also noted between radon progeny and cigarette smoke. Possible confounding factors in this study were asbestos, arsenic, asbestiform materials, chromium, nickel, and diesel exhaust.

3. Iron Ore Miners in Kiruna, Sweden. There were 41 lung cancer deaths among this cohort, of which only 13 were among underground miners and 4.5 were expected for the group (Jorgensen 1973). The data were collected from death registers in the community, and the age distribution and smoking histories were not adjusted, although 12 of the 13 were smokers. The radon exposure problem was not realized until 1970, when the concentration of radon was first measured and found to be in the range of 10–30 pCi/L. Histology was performed on only 7 of the 13; one was found to be of the oat cell type.

Among a similar cohort for both Kiruna and Gillivare, a case-control study involved 31 lung cancers, and although it showed an effect of radon levels causing lung cancer, the numbers were too small. The effect of smoking appears important, but this study can only be regarded as suggestive (Damber and Larsson 1982).

4. Iron Ore Miners in the UK. Examining 5,811 miners' death certificates issued between 1948 and 1967, from the communities of Whitehaven and Ennerdale, UK, showed 36 lung cancer deaths compared to 20.6 expected for non-miners locally or 21.5 nationally (Boyd et al. 1970). Many of the miners had previous experience, although there were no individual exposure or smoking records. They concluded that iron miners working underground suffered a 75% increase in mortality from lung cancer. In 1981 only 4% of non-coal miners in the UK were exposed to levels exceeding 4 WLM/yr (O'Riordan 1981; Dixon et al. 1985).

5. Metal Miners in the US. Forty-seven lung cancer deaths were registered here with 16.1 expected, yielding a SMR of 292 and $p < 0.01$ (Wagoner et al. 1963). Possible confounding factors for this study include sulfur, iron, copper, zinc, manganese, lead, arsenic, calcium, fluoride, antimony, silver, and nickel, in order of diminishing concentration.

6. Navajo Uranium Miners in the US. This study involved 32 lung cancer cases where only 23 were uranium miners, based on death certificates and some histology data (Samet 1984). The American Indians have lower rates of lung cancer than other US populations, which is likely due to less cigarette smoking. Only 5 of the 23 miners smoked as much as 4–8 cigarettes/d. The result of this study is "highly suggestive, but inconclusive due to incomplete and inconsistent ascertainment of occupational histories."

7. Norwegian Niobian Miners. This study involved 318 of whom 77 were underground miners (Solli et al. 1985). In this cohort 9 lung cancers were observed compared to 0.8 expected. The radon progeny levels were in the range of 1–4 WL, while the thoron progeny levels were in the range 0.2–0.4 WL. The mine only operated from 1951 to 1965, and the follow-up period was from 1953 to 1981. Some questions still remain about the exposure measurements, since they were made during a 2-d period in the winter. Not all of the miners could be located for follow-up. Dust levels were high which could have caused self-absorption on the measuring filters, suggesting that the exposure was really higher.

8. Norwegian Magnitite Miners. This small study involved 332 underground miners (Leira et al. 1986). The workers were in the mines between 1940 and 1960, for periods of more than 36 mon. The ratio of observed to expected for lung cancer incidence was 1:1. The data used were from the Central Bureau of Statistics. The study was too small and did not show any increases of lung cancer incidence due to low alpha radiation. The observed to expected ratio at the Fosdalen site was 2.8, but the 95% confidence range was 0.2–3.1. Thus the range includes 1.0.

9. South African Gold and Uranium Miners. This cohort involved 20,000 white and 250,000 black miners. The exposure levels were generally low because of the low average uranium grade, and the levels were generally below 0.3 WL (Steinhausler 1988). Twelve mines were involved and the exposure levels were "guesstimates." It was concluded that this study could not be used in estimating risk quantitatively because of lack of quality assurance on histopathology, the thoron levels, the presence of other non-radioactive hazardous agents, and a poor level of follow-up.

10. Tin Miners in Yunnan, China. Among 12,243 tin miners examined between 1975 and 1981, there were 433 cases of lung cancer compared to 29.8 expected. The SMR was 1451 and the p value was not specified but estimated to be < 0.05 (NIOSH 1987). The mines are known to have arsenic levels, but the time when workers started, the number of workers lost in follow-up and any adjustments made for age and smoking were not specified. This is an ongoing study and will be discussed later.

11. Uranium Miners in France. This cohort involved those who had worked at least 3 mon in underground mines between 1947 and 1972 (Tirmarche 1985). There were 36 lung cancer deaths compared to 18.77 expected for a SMR of 191 where $p = 0.002$. This study did have individual radon exposure data.

12. Zinc-Lead Miners in Sweden. Among the cohort for zinc-lead miners in Sweden, there were 29 lung cancers of which 21 were underground miners (Axelson and Sundell 1978). They claimed a 16-fold increase in risk but failed to match for age and smoking history and did not obtain individual smoking histories.

13. Other Exposures. The association between occupation and lung cancer risk was examined in a population-based, case-control study of 506 patients and 771 controls in New Mexico (Lerchen et al. 1987). For females, lung cancer was not found to be associated with employment history, but the power of the study was limited. For males, elevated risks were found for the uranium mining industry (odds ratio 1:9), underground miners (odds ratio 2:1), and welders (odds ratio 3:2). Potential limitations for this study include formation bias, selection bias, and inadequate statistical power for effects of a relevant magnitude.

Worldwide, water and air containing elevated levels of radon have been used by parts of the medical profession as a therapeutic agent for bathing and inhalation in Austria, Bulgaria, Czechoslovakia, France, Germany (East and West), Italy, Japan, United States, and USSR (Steinhausler 1985, 1988). In spas the workers receive higher doses than do the individual patients and thus they are at increased risk (Uzunov 1981). But due to the small numbers involved, their mobility and lack of individual exposure levels, statistically significant epidemiology studies are not possible. Other studies provide exposure data in surrounding spas, but none are complete enough or involve enough people for a statistically significant epidemiology study (Clemente et al. 1984; Pohl-Ruling and Scheminzky 1972; Pohl-Ruling et al. 1982; Pohl-Ruling and Fisher 1983).

Some preliminary studies have been conducted of the epidemiology due to exposure to radon and its progeny in homes. One such study is being conducted by Argonne National Laboratory in the Reading Prong area of Pennsylvania. This study involves 500 cases and 500 controls and is restricted to Berks and Lebanon Counties. This limitation has been decided due to the difficulty in obtaining direct data on the radon levels and histories of the inhabitants. The study is envisioned to last several years. Another study currently being conducted in the US is an addition to a study in New Jersey by the National Cancer Institute (NCI) of lung cancer in women. This study is limited to women who have only lived in one residence.

A study has been conducted in Stockholm, Sweden, involving 292 female lung cancer cases (Svensson et al. 1987). The exposure level was 55.0 ± 49.2 Bq/m^3 EER.[1] The cases were diagnosed between 1972 and 1980, as oat-cell and other types of anaplastic pulmonary carcinomas. The relative risk was 2.2 ($p = 0.01$), however, when the error range is included, a relative risk of 1.0 is easily within the range of error. Thus the study is not definitive regarding a possible effect.

---

[1]EER is the Equilibrium Equivalent Concentration of Radon. This is, for a non-equilibrium mixture of short-lived radon progeny in air, that activity concentration of radon in radioactive equilibrium with its short-lived progeny which has the same potential alpha-energy concentration as the actual non-equilibrium mixture).

Other preliminary studies of exposure to radon in residences include Port Hope, Ontario (Lees et al. 1987), and Sweden (Damber and Larsson 1987; Stranden, 1986). These are preliminary studies with no specific conclusions.

In Badgestein, Austria, there are 300 hotels, 14,000 beds and approximately a million thermal baths given per year. The indoor radon activities are of the order of 10 pCi/L of radon-222. Over the years "positive effects" have been identified for all the diseases of the rheumatic group, the vascular diseases, disorder of the endricine organs and metabolic disorders (e.g. gout) as well as for gerontal complaints (Pohl-Ruling and Scheminzky 1972; Stinhausler and Pohl 1973; Pohl et al. 1977).

A comparison of the SMR's and excess lung cancer cases/10^6 person-years WLM between several of the epidemiology studies discussed here are tabulated in Table 3.

In a study of US coal miners, the SMRs for lung cancer showed great variability (Rockette 1977, 1980). The SMRs are in the range of 59–192. It was concluded that the excesses seen are well within the range of what can be explained by differences in smoking habits and more studies are needed to be conclusive.

III. General Comments on Epidemiology Studies

The discussion here is based on analyses in the literature and on conversations with many of those directly involved in the miner studies and analysis and development of the data for health effects on radon exposure.[2] Some of this is based on scientific judgment, intuition, and best guesses. As is indicated, there is consensus on some questions and not on others.

A. Exposure

Perhaps the most universal problem area with the existing epidemiology studies for the relationship between radon and lung cancer is the area of exposure estimates and measurements. In many cases the exposure data are based on estimates that use measurements in nearby mines or involve measurements at only a few times during the year. One problem is, in order to do monitoring, permission of the mine owner is needed. If the mine owner thinks there is a problem he may keep the inspectors waiting while he makes changes in the ventilation system or mining locations to make the conditions appear better.

One important factor in determining exposure levels for doing dose–response studies is whether the measurements have been made of radon (in pCi/L) or of its

---

[2]The following discussion is based primarily on interviews with the following people: V. Archer, B. Cohen, F. Cross, W. Ellett, A. Goodwin, N. Harley, R. Hornung, G. Howe, Z. Hrubec, J. Johnson, C. Land, E. Laterneau, J. Lubin, F. Lundin, W. Mills, J. Muller, N. Nelson, J. Puskin, R. Roscoe, G. Sacamano, J. Samet, J. Stebbins, R. Toohey, J. Wagoner, and M. Wrenn.

progeny (using units of the WL). These two quantities are related by the equilibrium level of the progeny, which in general is not 100% for all progeny. In order to translate radon measurements into progeny values, assumptions have to be made about the equilibrium level, and this depends heavily on ventilation rates. However, much is known about variation in equilibrium rates, and this effect is seen by most as a small possible contribution to the overall uncertainty.

An additional contribution to risk in the environment due to radon can come from the extra exposure to thoron. Few measurements exist for thoron, although it was looked for in the US miner studies and not found. In most cases the dominant contribution to dose and risk is from high linear energy transfer (LET) radiation from alpha emitting progeny of radon. However in some cases where the Th-232:Ra-226 ratio is high, inhaled thoron progeny may be important. The thoron progeny are, in these cases, more important when the mine air is young (Stranden 1985). Their contribution to overall exposure and risk is expected to be small, since it has a short half-life compared to radon (1 min vs 4 d). It has been estimated that the contribution of thoron to the overall toxicity is of the order of 5–10%.

Another contribution to uncertainty in exposure estimates is due to the difference between the dust in mines and that in residences. There are several variables involved here, including breathing rate, which is composed of respiratory frequency and tidal volume, oral vs nasal inhalation, dust particle size, the unattached fraction, amount of cigarette smoke present, and other particulates, such as those from a kerosene heater or stove. Most investigators saw this as a complex issue but felt that the different contributions canceled each other and that the overall differences were less than a factor or two. The biggest differences were due to cigarette smoke and the improved ventilation rates in modern mines.

Some recent exposure estimates have been based on whole body measurements. This procedure is quite good and is based on the ^{210}Pb in the skeleton which has a long half-life, and is a possible indicator of exposure to radon progeny.

B. Possible Interaction Between Smoking and Radon Exposure

Perhaps the most important variable or complicating factor in the determination of the risk due to exposure to radon outside the obvious ones of dose and time, is the effect of the simultaneous exposure to cigarette smoke, either actively or passively. This single question can draw more conversation amongst those involved in epidemiological studies of the effect of radon than any other single subject. In spite of the controversy there is a general agreement that the effect of cigarette smoke is more than additive and less than multiplicative in its relation to exposure to radon. There are those who are influenced by the data from the atomic bomb survivors that show this relationship to be additive. The opponents, however, argue that the atomic bomb survivors were exposed primarily to low

LET radiation, and the effect is different for the high LET radiations from alpha particles emitted by radon progeny. There are also those who think that the data are too weak to make a judgment and say that it is anybody's guess what the relationship really is. This latter group would contend that the data are too noisy, and there are too few cases in each risk and exposure category to be able to tell if the relationship is multiplicative. In general the age and time dependency of the data tend to cluster and make analysis difficult. The relationship has not yet been determined for animals. There is enough variation in the data for some to conclude that it is not inconsistent with a supramultiplicative interaction.

In order to analyze the data the relative risk function can be expressed generally as:

$$R = 1 + b_1 \text{ WLM} + b_2 \text{ WLM}$$

for an additive relationship in which b_1 and b_2 are parameters or specifically,

$$R = (1 + 0.31 \times 10^{-2} \text{ WLM})(1 + 0.51 \times 10^{-3} \text{ packs})$$

for a multiplicative relationship where packs are the total cigarette packs smoked (Whittemore 1983). The parameters in the second equation were derived using the Colorado miner data for which the exposure rate used was extremely high and has been suggested to be overestimated. They concluded that there is a synergistic effect, where smokers with 20 pack years of cigarettes experience radiation-induced lung cancer rates per WLM that are roughly 5 times those of non-smoking miners.

From the analyses of the early 1960s it was thought that the interaction between inhaled radon progeny and cigarette smoke was multiplicative. However, it has been determined since then that smoking causes the lung cancer to occur earlier, making the interaction less than multiplicative. For most of the epidemiological studies, whether of miners or general populations, the data are weak enough so that one cannot distinguish between an additive or multiplicative interaction.

Whether the effect of smoking is additive or multiplicative is related to the mechanism by which the cancer occurs. If cigarette smoke is a promoter in this relationship then one would expect the relationship to be between additive and multiplicative. A somewhat simplified picture can given an idea of what is involved. Using the idea of a multistage model, if two agents act at the same stage one would expect the effect to be additive. However, if the two agents act at different stages the effect is multiplicative. It is suspected that smoking acts at an early and late stage, and that radon acts at an early stage; thus it is expected that the combined effect is in between an additive and multiplicative relationship. Because of this relationship, it is likely that the effect will appear to be more multiplicative in the early years.

One of the problems met in trying to unravel this question is that many miners are heavy smokers. It is thought that if only nonsmokers were studied by comparison, the effect would be found not to be multiplicative. For example the relative risk is ten times greater for non-smoking miners than for smoking miners, however, these data are based only on preliminary studies.

At the other end of the spectrum of opinion, it is possible that of the estimated 100,000 lung cancer deaths occurring annually from cigarette smoking, some may really be due to radon. This could be due to the effect of smoke providing a site for radon progeny to adhere, however, others have pointed out that smoking causes an increased flow of mucus which protects the lung from exposure to alpha radiation from radon progeny. Using data from US uranium miners, the hypothesis is postulated, on the basis of both animal and human data, that in the absence of cigarette smoking, long latent periods are present. However, in the presence of cigarette tar, the radiation-induced cancers will appear at an earlier date (Archer 1984).

It has been suggested that radon progeny have a direct relationship to the mechanism by which cigarette smoke causes lung cancer (Martel, 1975, 1983, 1987; Maronen et al. 1987). It has been observed that the presence of cigarette smoke in indoor air almost doubles the radon progeny concentrations (Bergman and Axelson 1983; Bergman et al. 1986).

An important factor in this relationship is the alpha-radiation dose from inhaled ^{210}Po deposited in small volumes ("hot spots") of bronchial tissue. Because most smoke particles are soluble in lung fluid, the bulk of the inhaled particles and ^{210}Po are cleared from the deposition sites in the lung. The strength of the localized particles in the lung range in activity from 10^{-6} to more than 10^{-4} pCi. Thus the small volume around the particle is subjected to alpha interaction rates between 100 and 10,000 times the natural level. Smoking enhances the number of submicron particles in the air to which radon progeny can adsorb. It has been estimated that a radiation dose of 80 to 100 rads is delivered at the bifurcation to approximately 10^7 cells (Martel 1983). It is further suggested that the process involves two stages. The beta radiation plays a primary role in the initiation and progression of the malignancy, and the alpha radiation serves mainly as the promoter.

Perhaps it is safe to conclude at this time that the general consensus of the scientific community is that cigarette smoking and exposure to radon are somewhere between a submultiplicative and a supramultiplicative relationship. The reason for the uncertainty is the lack of statistically significant data. The demands of statistical accuracy may be such that we may never know more than this. The actual relationship will remain a transcientific question, one that can be asked but likely not completely and precisely answered. The smoking interaction may not be better defined in the next 10 to 20 years, and by that time cigarette smoking may have worked itself out of the population. It will likely start its reduction in the upper classes and in time move out of the lower classes. The cigarette companies may be virtually out of business long before the possible interaction of radon and cigarette smoke is clear.

C. Other Variables

As discussed in more detail in the section on risk models, there is an effect on risk rate as a function of age. There is a dropping-off of risk rate after a time of the latent period following the last exposure. Besides these reasonably well understood variables, there are others for which less is known, or about which more needs to be determined. These other variables include: censoring of those entering the cohort because of previous experience in mines, censoring of those at higher ages due to losing track of them, possible different sensitivities of women and children, and different dose rates as a function of dose and exposures to other contaminants such as asbestos, arsenic, chromium, nickel, or diesel fumes.

For many of the miner studies those included in the cohort have been exposed to radon in earlier mining experience. However, in several studies cases were rejected if they had previous mining experience. In general this problem is only really important for the lowest exposure group, and the general opinion of those in the field is that it could make a difference of less than a factor of two for this group. The overall contribution of previous experience to the risk rates estimated from miner studies is estimated to be less than 10%. For both the Colorado and Saskatchewan studies the contribution of previous experience in mines made no statistical difference in the estimated risk rates.

The follow-up for the major mining studies has been quite good with estimates of 98 to 99% of the miners being followed in most studies until death, except for the Czechoslovakian miners for which no information is available. This effect would be most important for the highly exposed persons and those that move because they develop lung cancer. The follow-up of all cases was estimated to be a very minor problem and cause a small uncertainty in the estimated risk rates.

There are few data concerning the effect of radon exposure on women or children, since the miner studies are for adult males. Thus there are no direct data concerning the different sensitivities for children or women. Indirectly, it can be noted that there appears to be no sex variable in smoking effects, however, if a difference would exist it likely would be related to hormonal differences. Although the animal studies are not complete, the preliminary results appear to support no difference between males and females, and there is no strong biological reason to expect a difference in either humans or animals. Some preliminary studies in the general population for non-smokers seem to show no difference between the sexes for radon exposure. It is possible that there could be a difference, if males were exposed to more initiators, and thus the exposure patterns will be important in any studies of this kind.

Children may have different sensitivities due to developmental causes, although this possibility is not clear. Because of the latent period, there will be no lung cancer for younger children, and the longer they are removed from exposure the more the risk decreases. It is possible that children have repair mechanisms that reduce or prevent lung cancer.

To answer the question of possible differences between males, females and children, a nationwide epidemiological study is needed to definitively show if the miner experiences can be extrapolated to residences directly. Other variables may be important such as ethnicity, skin color, and life styles.

There are several other environmental contaminants such as asbestos, arsenic, chromium, nickel, diesel exhaust and silica in mines that could contribute to health effects of miners. In general these are important to some degree, but likely and mainly to the low dose group. However, as atmospheres in mines improve with better safety and ventilation, the potential confounding with other contaminants becomes more important. Also many do not involve lung cancer or even fatal effects. Thus in the gathering of data with epidemiology studies, often the potential effects of other contaminants are missed. From the modelers point of view these effects represent a background that would exist, if there were zero radon exposure. This background can be corrected statistically in the data analysis. Also it is a rough approximation that the levels of these contaminants will, in increases and decreases, parallel those of radon. The consensus amongst those that study miners is that the effect of these contaminants is somewhere between relatively important and unimportant and something less than a factor of two. Most think that these effects vanish when all the different studies are combined since the levels of each in the different mines varies widely. Some think that it takes a lot of confounding to influence a risk estimate and that the emphasis on confounding, especially in this cases, is overstated. Perhaps the most this effect could be is to provide a variable to the radon risk factors, in the sense that the risk of lung cancer from radon exposure is very small and that the influence of these other factors might possibly be noticeable.

Another way to look at the influence of other contaminants is to consider another category in addition to those of initiator and promotor, that of a facilitator. A facilitator is a non-carcinogen that has an impact or is an affect that can shut down a natural defense mechanism. For example if exposure to a contaminant can shut down the mucus escalator motion, it can greatly "facilitate" the action of a carcinogen such as radon progeny. This is a mechanism that does not cause cancer, but which if it did not occur would possibly prevent it. Cigarette smoke could be an example of this phenomena in conjunction with radon progeny, since cigarette smoke has both a carcinogenic and a non-carcinogenic component. It is still an open question, if radon progeny is an initiator or promoter. More work is in progress on this question, especially for animals.

A common criticism of epidemiological studies and particularly those involving radon progeny is that the cohort is too young, and that the cohort has not been followed up long enough. Few would argue that one must wait until the last person has died or wait for the maximum lifetime of 110 yr. However, the existing cohorts are still aging and there is a time after exposure, when the contribution by waiting longer is very small. Sufficient time needs to elapse for follow-up to be sure that most of the cases have been determined. There is a general consensus that when 75% to 80% of the cohort are dead, or when 20 to 40 yr have elapsed since last exposure, that most of the data are in. The problem is that old age is important, since lung cancer reaches a peak for those in their 60s and 70s. Thus if no lung cancer has appeared by age 75, then it becomes less important with respect to other causes of death.

Some studies have suggested that as higher doses are experienced, the risk to exposure rate decreases. This raises questions, e.g., does it alternatively increase at lower rates? The answer is not known at this time, although there is some suggestion of such an effect in the US data. This relationship may never be known and is viewed by most as unlikely unless there is a threshold. The effect is seen only for very high doses such as over 1000 WLM. Biologically, one would expect the risk curve to curve downward at lower doses due to a possible repair mechanism and a threshold. Some think that there are repair mechanisms that produce a theshold. The general view is that the dose–response curve is linear except for this possible bending over at high doses. At very high doses it has been known for a long time that there is a decrease in risk rate due to cell killing. However, few believe that this occurs at the levels found in most mines. Some think that above 3000 cumulative WLMs the dose-response curve bends over, and this shows the effect that some call "wasted radiation." That is above a particular level of dose, the radiation causes less of an effect and thus in that sense is wasted. Since this effect has not been seen in all studies, there is suspicion of an inconsistency perhaps due to a dose overestimation or a shift of persons between categories. It seems clear that this effect, if it exists, should not be extrapolated to lower doses; because its biological origin is not clear, and it is an empirical observation that needs follow-up research to determine if it is valid, and if so what is the causal mechanism.

D. Histopathology

Some researchers have suggested that the cell types seen for lung cancer resulting from cigarette smoking and those resulting from exposure to radon progeny are different. In both these areas of study there are some differences, because of the way in which pathologists categorize cell types. However, the general consensus is that the mix of cell types, when viewed over a lifetime is roughly the same. The resulting cells are not distinguishably different, whether they are exposed to cigarette smoke or radon progeny. In the early years of miner studies, it was believed that radon progeny led to more small cell carcinomas, but it now appears that smoking causes cancers to appear earlier, and these earlier cells are more of the small cell type. It further now appears that as time goes on, the small cell type frequency decreases and the epidermoid cell frequency increases, while the frequency of adenocarcinomas is unchanged in time. It is also believed that oat cell carcinoma occurs earlier. The relative frequency of cell types is similar in animals to that in humans, except that in humans the K cell is the progenitor of oat cell. One possibly complicating factor is that death certificates often are used, and although lung cancer is rarely misclassified, it is possible that the cancer had metasticized from another site.

The specific histopathological cell type caused by radon progeny has over the years been controversial (NAS 1988, page 497). It was thought that there was only one kind of cell such as the oat cell, but not all kinds were reported. The identification of specific cell type is not consistent over the years, nor between different countries or studies. The World Health Organization (WHO 1982) lists eight types of malignant lung tumors. They are squamous (or epidermoid) carcinoma, small cell carcinoma, adenocarcinoma, large cell carcinoma, adenosquamous carcinoma, carcinoid tumor, bronchial gland carcinomas and others (which include soft tissue tumors, mesothelial tumors and unclassified tumors). Of the first four of these categories, the percentage found in US male populations is 35, 17, 25, and 9% respectively. Thus the first four categories represent 86% of the total. Studies of uranium miners shows the following percentages for smokers: 35.96, 32.30, 13.20 and 11.80%, and that for nonsmokers as: 40, 24, 12, and 20% (Saccomanno et al. 1988).

It has been suggested that the site for the lung cancers due to radon progeny is the bifurcations or the carina where one passageway splits into two. This is logically where one would expect dust particles carried in the air in the passageways to lodge. However, it is not clear where most tumors start, because the cancerous tissue is large enough by the time it is found to make its specific origin unclear. Most of the data are from death certificates, and therefore no information exists about tumor location. In general some tumors are found in the area of the bifurcation, and some are found down the passageways.

E. Other Health Endpoints

There have been some marginal suggestions that other health endpoints result from exposure to radon progeny. The most likely suggested are those for skin cancer and stomach cancer. The skin cancer found is primarily the non-fatal kind which would not be detected in studies looking only for mortality. The Czechoslovakian studies report skin cancer with a SMR = 453 ($p < 0.05$) (Secova et al. 1978). It is not likely that alpha particles would cause skin cancer, since the alpha particles are not energetic enough to reach the sensitive cells in the skin. However, there is a possibility that there may be an effect from gamma rays. The data in this area are usually not statistically significant and sparse. The general opinion is that it is a small but unlikely effect and may be due to the arsenic in the mines. Estimates have been made of the deposition velocity of Aitken particles carrying radon progeny on hairy surfaces (Martel 1982; Martel and Poet 1982). The 7.7 MeV alpha disintegration from ^{214}Po can possibly explain the high incidences of skin cancer.

The stomach and upper gastrointestinal track is a likely place to look for other effects of inhaled radon progeny, since that is where the mucus can move from the lung. Stomach cancers have been reported in several studies as follows: tin miners in Cornwall, England, showing SMR = 200 (Fox et al. 1981), Ontario gold miners with SMR = 148 (Muller et al. 1985), US metal miners with SMR = 149 (Wagoner et al. 1963) and Swedish iron miners with SMR = 189 (Radford and Renauld 1984).

One other health related endpoint that definitely is related to radon exposure is chromosome damage. Inhalation in spas of up to 40 pCi/L of radon did not produce any detectable increase of aneuploidy or chromosome breaks (Pohl-Ruling and Schemizky 1972). The total chromosome type aberration frequency (sum of numbers of dicentrics, rings, terminal and interstitial deletions) shows that very low doses result in a sharp increase of around a few tenths mGy/mon of alpha and gamma blood dose with plateaus in the range of 0.5 to 1.0 mGy/mon (Pohl-Ruling and Fisher 1983). It has been hypothesized that after some damage to the DNA, additional repair enzymes are produced to control the radiation damage (Pohl-Ruling et al. 1977). It is not clear what this information means for human health, and it cannot be related to lung cancer. Thus chromosome damage can only be regarded as a biological indication of exposure to ionizing radiation.

Other potential health related effects due to radon exposure include metabolic changes as indicated by alteration of cellular oxygen consumption and alterations of membrane properties as measured by the transmembrane resting potential (Reubel et al. 1987).

Muller (1967) reported a statistically significant decrease in the sex ratios of live-born children of 0.383 for women exposed to ionizing radiation in uranium mines. A similar study of men showed no statistically different sex ratio. The normal ratio is 0.515. Other work in this area shows that it is not completely clear that these anomalies are due to radiation, and that they may not be due to a single cause (Wiese 1981a, 1981b; Wiese et al. 1986).

F. Threshold?

There is no statistically defensible evidence that a threshold does or does not exist for ionizing radiation in general and for exposure to radon specifically. There have been many suggestions of a threshold over the years, and some today believe it exists.

One kind of a threshold is the "practical theshold." In this situation it is noted that for lower doses on a dose–response curve, the latent period increases. Eventually a dose is reached where the latent period is longer than the lifetime. Perhaps the most direct phenomenon that would indicate the existence of a threshold is a repair mechanism. It is well known that lung cells have a good repairability. The argument then goes that by natural selection and evolution, we have adapted to the background ionizing radiation, and to survive have developed repair mechanisms to cope with radiation. Some have even put numbers on possible thresholds, although most are in the range of a few hundred WLMs. Some are suspicious that a threshold exists above the levels found in residences and even above the level of 4 pCi/L (4 pCi/L corresponds roughly to 0.02 WL which for a 70 year lifetime would be an exposure of approximately 70 cumulative WLMs/lifetime).

G. Future

The best hopes for future information being developed that may shed some light on the questions discussed here include: updating and re-examining existing studies, conducting studies in residential areas and the evolving data from the Chinese tin miners. There are two studies being done with the Chinese. One involves a cohort study of 22,000 workers with over 1000 lung cancer cases. The other is a case-control study with 600 cases and 800 controls. These two involve a population that is not very mobile and includes children as young as age 12. This new study will hopefully provide some new insights into the area of health effects of radon exposure. Of the epidemiological studies under way, the ones in Reading Prong and New Jersey will, for the first time, provide information concerning residential exposures.

IV. Dosimetry

The epidemiological studies of miners demonstrate clearly that exposure to radon progeny are correlated with lung cancer. However, they do not demonstrate a cause and effect relationship. In order to examine this question more completely, an effort is taken to determine some aspect or property of the miners' experience which might be labeled as a cause (Crawford-Brown, in Cothern and Smith 1987, NCRP 1984b; James 1988). One important piece of information is the determination of the dose equivalent delivered to the cells responsible for lung cancer. This cannot be measured directly and must be estimated.

The dose equivalent delivered to the lung is highly nonuniform, and thus must be examined at the cellular level. The field that has developed to pursue this direction is known as microdosimetry. Its objective is to determine the specific energy delivered to small volumes of the order of one μm in diameter or less. A dosimetry model is a collection of mathematical functions used to calculate absorbed dose. The dose delivered by radon progeny to the lung depends on both the aerosol involved and the physiology of the lung.

The computation of the dose equivalent delivered to cellular subpopulations is made in several steps:

1. The first is to determine where the inhaled progeny are deposited, which is determined by the size of the aerosol particles, the volume of air inhaled and the age of the individual.

2. Next, the movement of the progeny is determined along the mucociliary blanket.

3. Then it can be estimated where the individual atoms decay.

4. Finally, depth-dose curves are developed describing the dose equivalent delivered to the cells.

A. Anatomy of the Lung

Generally the lung is modeled in regions of nasopharyngeal (NP), tracheobronchial (TB) and pulmonary (P). The description of the region beyond the nasopharyngeal is of a series of bifurcating tubes or passageways. Each passageway splits to produce two or more passageways with the split resulting in new passageways whose diameters are smaller than those of the original.

The surface of these passageways consists of pseudostratified columnar epithelial cells coated by cilia. This layer lies above the basal cells which rest on a thick basement lamina. The basal cells are the source of new epithelial cells.

The branching scheme of the lung is modeled in various ways which include asymmetry in numerous generations. In the Wiebel A model, for example, 24 generations are used with the number of branches increasing to 300,000,000 in the last generation. The Weibel A model involves dichotomous branching, whereas the Weibel B model does not.

One of the more important variables is age and unfortunately little information is available to describe how the lung changes with age. However, from the available information it appears that changes are significant up to about 16 years of age.

B. Properties of Alpha Particles

A number of characteristics of alpha particles are well known and directly applicable to microdosimetry including scattering properties, stopping power, linear energy transfer, range, straggling and radiobiological effectiveness. In general alpha particles travel in straight lines, however, small deflections do occur which lead to straggling or random variations in the range. The alpha particle looses very little energy in each individual interaction, and at low energies it takes on two electrons and becomes a helium atom. The stopping power for alpha particles of interest here is two to three orders of magnitude greater than that of electrons and is in the range of 100 kev/μm. Linear energy transfer gives the rate at which energy is laid down close to the particle track. The range of alpha particles is fairly sharply defined and for energies of a few MeV is in the range of a few to 100 μm. The relative biological effectiveness is a quantity used to compare radiations, since in general they do not always produce the same effect for the same energy deposited.

C. Modeling of Deposition and Dose Equivalent

The processes involved in deposition included impaction, sedimentation and diffusion with some minor contribution from deposition of charged particles by the image force resulting from the rearrangement of charge on the surface of lung passages. The relative importance of the processes depends on the size distribution of the inhaled aerosols and the unattached fraction.

The dose equivalent for a subpopulation of lung cells is determined assuming that alpha particles travel in fairly straight lines due to their large mass. Knowledge of stopping-power allows estimates to be made of depth–dose curves. These give the dose equivalent at various depths as a function of the energy of the alpha particles and the cell structure.

It is assumed that the critical cells are those which have not differentiated and are still capable of differentiation and division. These are the cells of the bronchial epithelium which are replaced by division of basal cells, suggesting that the basal cells should be regarded as the most radiosensitive, although this is not entirely clear at this time. Lacking more detailed information basal cells are considered the critical ones. Recently it has been argued that the average dose to all epithelial cells should be considered (James 1988).

D. Factors Involved in Dose Calculations

The calculation of dose equivalent to the lung depends on a number of specific variables including:

Physical Characteristics
  fraction of unattached ^{218}Po
  progeny equilibrium
  particle deposition
  particle size distribution
  method of computing dose
  flow (turbulent or laminar)
  humidity

Biological Characteristics
  breathing pattern
  bronchial morphology
  mucociliary clearance rates
  location of target cells
  mucus thickness

There are three versions of the model that have been used in the dosimetry estimates for radon progeny in the lung. These were developed by Harley and Pasternack (1981), Jacobi and Eisfeld and James and colleagues (James 1988). These differ in some important characteristics such as clearance mechanisms, cells at risk and lung morphology, however, the estimates by them for dose per unit exposure for specific conditions in homes and mines differ by a factor of three at most (NAS 1988).

The version developed by Harley and Pasternack involves clearances by mucus only, and uptake in the blood is ignored. It assumes that the only cells at risk are the basal cells of the tracheobronchial epithelium. The dose is calculated at a fixed depth of 22 µm below the surface for the first ten generations and at 10 µm beyond the tenth generation.

That developed by Jacobi and Eisfeld uses a symmetrical lung model and considers the effect of variable depth of target cells. Clearance is by both mucus and the blood.

And that developed by James and associates assumes a uniform probability distribution of depths for target cells. It includes the dose to epithelial tissue as dissolved activity being transported through the bronchial membranes into blood.

Studies have been made of the ciliary streaming (Hilding 1957) and the volume of the bronchial tree (Hilding and Hilding 1948). These studies show that there is a progressive increase of from 50 to 100 percent in volume of the tracheobronchial tree from larynx to bronchi. The only effective air cleaning mechanism found was direct impingement from air on the mucus blanket lining the tree. In the passageways the ciliary stream encounters a number of naturally occurring obstructions such as squamous islands, bronchial openings and vocal cords. The ciliary stream may be retarded or enhanced by cigarette smoke or other air pollutants which slow or temporarily stop the movement. This facilitates prolonged action of any carcinogen in the ciliary stream at these spots (Hilding 1963).

Using the hollow case model of the tracheobronchial tree it has been shown that the flow of particles is turbulent and that the laminar flow assumption is not a good one, since it gives erroneous estimates for deposition at bifurcations (Cohen 1987; Martonen et al. 1987). This may be another example of chaotic behavior in nature (Gleick 1987).

From the existing data it does not appear that the characteristics of aerosols in mines and residences is very different from those involved in the transport of radon progeny to the lung (NAS 1988). An earlier study documented in some detail the range involved (Guimond 1979). The four important characteristics are the median diameter of the aerosols, the concentration of the aerosols, the unattached fraction, and the equilibrium factor. The median aerosol diameter in residences is in the range of 0.1 to 0.2 µm, while that in the mines is 0.17 µm and that outdoors is in the range 0.04 to 0.3 µm. The largest range of concentrations was between 10^4 and 10^5 particles/cm^3 in residences and outdoors, while there was a range of 10^3 to 10^6 particles/cm^3 in mines. The origin of these dusts varies, and if cigarette smoke is present the concentrations can be as high as 10^8 particles/cm^3. The unattached fraction is 0.07 in residences, 0.08 outdoors and 0.04 in mines. The equilibrium factor varied but was similar for residences, mines and outdoors.

A range of dose conversion factors has resulted from these model calculations and falls in the range from 0.2 to 10 rad/WLM (2–100 mGy/WLM) (United Nations 1982). Many individual estimates have been made and fall usually at the lower end of this range; 0.2 to 2 rad/WLM (Harley and Pasternack 1981), 0.3 to 1.0 rad/WLM (James 1988) and 0.6 rad/WLM (Jacobi 1985). It has been observed that the dose per unit exposure is essentially invariant over a wide range of conditions allowing miner data to be directly extrapolated to residential estimates (Harley 1984b).

E. Dose Equivalent Estimates

A number of researchers have estimated the dose equivalent to the lung (Harley and Pasternack 1981, and numerous papers referred to in NCRP 1984b; Cothern and Smith 1987; Harley 1987; James 1988; NAS 1988). The estimates vary over a substantial range and from about 35 mGy/WLM to about 500 mGy/WLM. It has been suggested (James 1988) that most estimates of the dose conversion factor for unattached radon progeny are in the range of 100 to 200 mGy/WLM. It has been shown that over a wide range, the lung dose is proportional to the concentration of radon gas (James 1988).

Criticisms of dosimetric models include not using physiological models, not involving the effect of smoking or results of animal experiments and focusing on the wrong cells (Martell 1987). The effect of smoking appears to be between additive and multiplicative, and this needs to be considered in any discussion of lung cancer by radon progeny. Also the difference in rates between sexes must be included. The difference between animal experiments and epidemiological studies needs to be more carefully involved in the analysis of dosimetric models. The assumption that the basal cells are the critical ones is also open to criticism.

V. Animal Studies

Since the 1950s, research has been conducted on the health effects of inhalation of radon by animals. The discussion here is brief, and more details can be found in the chapter on health effects in Cothern and Smith (1987). The laboratories involved in animal studies include two in the US, the University of Rochester and the Pacific Northwest Laboratories, and the Compagnie General des Matieres Nucleaires (COGEMA) in France.

Several biological effects have been observed in dogs and rodents following the inhalation of radon progeny; primarily, respiratory epithelial carcinoma, pulmonary fibrosis, pulmonary emphysema and lifespan-shortening. Lesions above the trachea (extrapulmonary lesions) and nonpulmonary lesions have been reported in these studies (Cross 1988).

In general in laboratory animals, the tumorogenic potential increases with the increase in exposure, the decrease in exposure rate and the increase in unattached fraction and disequilibrium.

The influence of cigarette smoke depends on the temporal sequence of the exposures. In beagle dogs, alternating cigarette-smoke and radon-progeny exposures on the same day produced a decrease in lung-tumor incidence from that produced by radon-progeny exposures alone. Smoke exposures completed before radon-progeny exposures did not alter the lung-cancer incidence in rats;

however, smoke exposures following completed radon-progeny exposures produced a synergistic effect.

Although lung-tumor data differ somewhat, the overall incidence data in adult male animals (rats, primarily) are similar to the present estimated lung-tumor-incidence data in man. The derived range in mean lifetime risk-coefficient for atmospheres with low percentages of unattached progeny, weekly exposure rates exceeding from 50 WLM, and data uncorrected for lifespan differences from control animals, is about 1 to 5×10^{-4}/WLM between exposures of 100 and 5000 WLM.

Some observational results have been reported in animals but not in humans, and this, of course, does not constitute proof of a difference. The increase in tumor production with increase in radon-progeny unattached fraction and disequilibrium, the importance of the temporal sequence of exposures to cigarette smoke and radon progeny, and the presence of extrapulmonary lesions, including carcinoma, were demonstrated.

Thus, radon decay product exposures are able to produce lung carcinomas in animals. It would appear that the dose-related coefficients in nonsmoking humans are lower than those in nonsmoking rats by perhaps a factor of two (Cross 1988).

VI. Risk Estimates

The analysis of risk due to the inhalation of radon progeny can be approached in two ways. The first, involving descriptive models, fits the data in terms of excess cancer as a function of age at observation for risk and age of exposure. Several groups have developed descriptive models based on the existing epidemiological studies, and the best known are described in some detail in section VIa. The second involves extrapolating the dose–response curves from epidemiology studies to residential exposure levels and estimating the total number of lung cancer fatalities expected in the US annually due to the presence of radon in indoor air environments.

A. Descriptive Models

It is appealing to try and understand the mechanism of cancer and the biological roots that cause it, however, the data are not complete enough to know how they vary over a wide range of exposures and ages. Some complexities in this analysis include temporal sequence of multiple exposure, varying dose rates and durations and possible synergistic and antagonistic interactions among multiple exposures. By contrast it is straight forword to develop descriptive models from the miner data for exposures to radon and its progeny.

In order to predict the individual risk due to low levels of radon progeny, models are needed to extrapolate the dose–response curve from the relatively wide range, for which data are available, to lower levels where environmental exposures occur. In general, several groups have attempted this procedure and have combined information about the epidemiology studies of the miner populations and the predictions of the dosimetric models. Depending on the relative importance given to the many variables, the weight given to different studies and the difference between the epidemiology studies and their results, the predictions of risk vary.

Some of the variables in these exptrapolations include: age at first exposure (increased risk for miners exposed at older ages), exposure rate (higher rate leads to lower attributable risk), ability of the lung to repair damage, blocking the data in different ranges (this effect can be eliminated by using a regression model such as Cox's 1972), latent period and the effect of smoking. It has been observed that a multiparameter model is needed to faithfully represent the temporal pattern of tumor appearance (Harley 1988).

Several recent groups have developed models to describe the extrapolation of risk including the National Council on Radiation Protection and Measurements (NCRP 1984b), the International Commission on Radiation Protection ICRP 1987), the Ontario miner experience and the National Academy of Sciences (NAS 1988 or BEIR IV). The characteristics of these different models were developed independently, and as will be shown, the models give significantly different predictions (see papers of NCRP Annual Meeting, June 1988, Harley 1988, Land 1988 and Lubin 1988). To give a direct comparison the notation of Land (1988) was used.

1. NCRP Model. The age specific mortality for lung cancer following a single radon exposure to X WLM at age t_o, given smoking history S, is;

$$H(t|t_o,X,S) = H_o(t|S) + 10^{-5} u(t) f(t - t_o) X$$

where $H_o(t|S)$ is the baseline disease rate for an individual not exposed to radon progeny and of age t with smoking history S, and

$$u(t) = 1 \text{ if } t > 40$$
$$= 0 \text{ if } t < 40$$

and

$$f(t - t_o) = \exp(-(t - t_o) \ln(2)/20) \text{ if } t - t_o > 5$$
$$= 0 \text{ if } t - t_o < 5$$

This model reflects an increased risk in miners first exposed to radon progeny at older ages such as over 40 (Harley 1988).

2. ICRP Model. The age specific mortality for lung cancer following a single radon exposure to X WLM at age t_o, given smoking history S, is;

$$H(t|t_o,X,S) = H_o(t|S) (1 + d(t - t_o) b(t_o) X)$$

where $H_o(t|S)$ is the baseline disease rate for an individual not exposed to radon progeny and of age t with smoking history S, and

$$d(t - t_o) = 1 \text{ if } t - t_o > 10$$
$$= 0 \text{ if } t - t_o < 10$$

and

$$b(t_o) = 0.0071 \text{ if } t_o > 20$$
$$= 0.0213 \text{ if } t_o < 20$$

About 91% of b corresponds to risk from ^{222}Rn, the remainder to ^{220}Rn.

3. *Ontario Model.* The age specific mortality for lung cancer following a single radon exposure to X WLM at age t_o, given smoking history S, is;

$$H(t|t_o,X,S) = H_o(t|S) (0.016 X_{5-9} + 0.034 X_{10-14} + 0.003 X_{15+})$$

where $H_o(t|S)$ is the baseline disease rate for an individual not exposed to radon progeny and of age t with smoking history S, and X_{5-9} is the exposure occurring during 5–9 yr prior to the calculation of risk and so forth.

4. *NAS or BEIR IV Model.* The BEIR IV model is based on miner epidemiology studies from the US miners in Colorado, the Swedish miners at Malmberget, the Ontario miners and the uranium miner data from Eldorado, Beaverlodge. The BEIR IV Committee first carried out separate analyses of the four cohorts and then analyzed the combined data. They used the Cox relative-risk regression model with a Poisson probability model for the number of deaths in each cell. The relative risk model was chosen because of the fairly general observation that in the cohort studies the excess risk increases markedly with age in a similar manner as does the background risk. What emerged is a time-since-sequence (TSE) model.

The age specific mortality for lung cancer following a single radon exposure to X WLM at age t_o, given smoking history S, is;

$$H(t|t_o,X,S) = H_o(t|S) (1 + 0.025 \, a(t) \, T(t - t_o) X)$$

where $H_o(t|S)$ is the baseline disease rate for an individual not exposed to radon progeny and of age t with smoking history S, and

$$a(t) = 1.2 \text{ if } t < 55$$
$$= 1.0 \text{ if } 55 < t < 65$$
$$= 0.4 \text{ if } t < 65$$

and

$$T(t - t_o) = 0 \text{ if } t - t_o < 5$$
$$= 1 \text{ if } 5 < t - t_o < 15$$
$$= 0.5 \text{ if } t - t_o > 15$$

The BEIR IV Committee concluded that the basic sampling variation could introduce a multiplicative standard error of 30% into the statistical analysis of the lung cancer risk coefficient. This 30% error corresponds to the 67% confidence level. Thus the uncertainty at a 95% confidence level (approximately plus or minus two standard errors) would be represented by multiplication and division by 1.7 (approximately the square of the standard error, 1.3). The Committee also concluded, "The imprecision that results from sampling variation can be readily quantified, but other sources of variation cannot be estimated in a quantitative fashion. Therefore, the committee chose not to combine the various uncertainties into a single numerical value." (NAS 1988, p. 42)

BEIR IV model includes both time-since-exposure and age at risk. Also they claim that the dose to the tracheobronchial epithelium from thoron progeny is, for an equal concentration of inhaled alpha energy, less by a factor of three than that due to the progeny of radon-222. The potential for lung cancer from thoron cannot be evaluated directly, because the exposure information from the miner epidemiology studies are almost exclusively for radon-222.

5. A Relative Risk Model Using Colorado Uranium Miner Data. Hornung and Meinhardt (NIOSH 1987) developed a quantitative relative risk model based on a cohort of 3366 white underground uranium miners from the Colorado plateau who worked at least one month in an underground mine. Updated to 1982, the cohort contains 255 lung cancer deaths. They used the Cox proportional hazards model which permitted the use of internal comparison groups while controlling simultaneously for such confounders as cigarette smoke, age and year of birth, and also time-dependent covariates such as cumulative exposure. They observed that miners receiving a given amount of cumulative exposure at lower rates for longer periods of time were at greater risk relative to those with the same cumulative exposure received at higher rates for a shorter time, thus the curve is convex (see Figure 4). They also observed that older miners are initially at greater risk when exposed than are those first exposed at younger ages.

A comparison of the predictions of the models discussed here is shown in Table 5.

As can be seen in Figures 4 and 5, the structures of the dose–response curves are quite different in shape and relative incidence at any given age of observation.

B. Calculation of the Population Risk

Following the format of the calculation used by the USEPA (1987a), the total US lung cancer deaths from indoor radon are a product of the following factors:

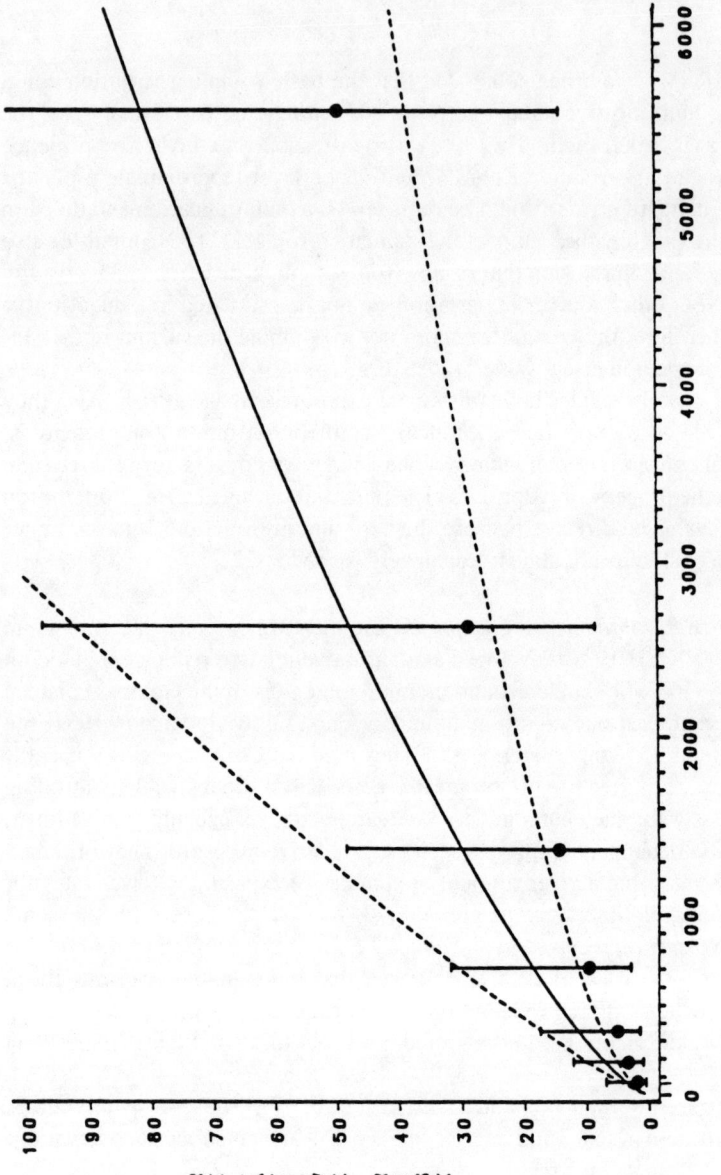

Fig. 4. Plot of the relative risk as a function of the cumulative radon progeny exposure (from NIOSH 1987).

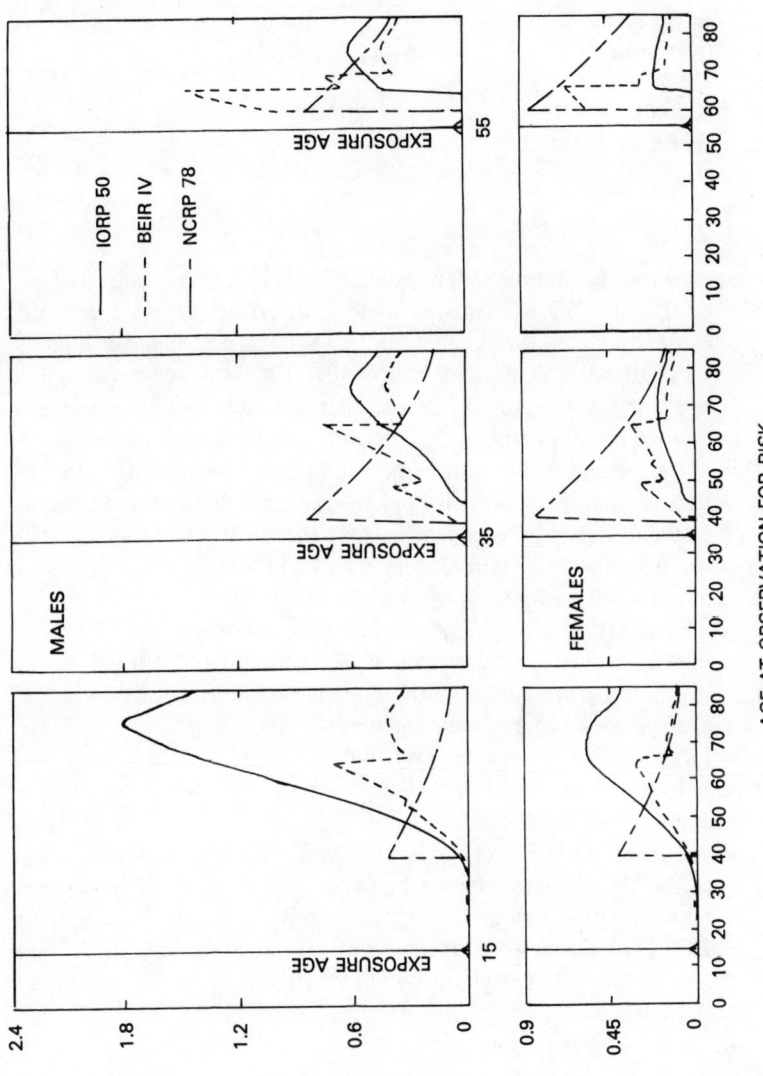

Fig. 5. Plots of the excess lung cancer incidence per 100,000/year as a function of age at observation for risk for the ICRP50, BEIR IV and NCRP 78 models (from NCRP 1988 with permission).

Table 5. Comparison of the Predictions of Descriptive Models

Study	Excess Lifetime Lung-Cancer Mortality (deaths/10^6 person WLM)
BEIR IV	350 (range 120–750)
NCRP 78	130 (range 100–200)
BEIR III	730
UNSCEAR 1977	200–450
ICRP	210
USEPA	190–550

C_R = average (mean) lifetime indoor radon decay product concentration
= 0.002–0.01 WL-life (this estimate is determined using the world estimate as the lower limit (UNSCEAR 1982) and an upper limit of a 2 pCi/L based on the data from Cohen (1987) that suggest the average is in the range of 1–2 pCi/L, that of Puskin and Yang gives an average of 0.9 pCi/L). A 50% equilibrium factor was used.

T = average interval of lifetime exposure in hr, following a 10-yr induction period during which no lung cancer will be observed, and assuming a range of occupancy of 50% to 90% and 73.88 yr life-span
= (0.50–0.90) (73.88–10) (365)(419,691.6) hr/life
= 3.236×10^5 to 5.825×10^5 hr/life

F_{WLM} = factor converting average cumulative indoor exposure in WL hr to working level months, accounting for 170 hr per month exposure period, and the difference in breathing rate between the average adult (15.3 L/min) and a miner (30 L/min).
= (1/170) (15.3/30) = 0.003 WLM/hr

R_{RRM} = relative lung cancer risk for lifetime exposure to radon, per WLM, using the relative risk model
= 1% to 3% per WLM (The strongest epidemiology studies are those of the US uranium miners, the Ontario and Czechoslovakian uranium miners which give risk factors in the range of about 0.9% to 2.3%. The recent work of Howe et al. (1987) gives a value of 3.28%)

TCR = underlying annual average of the US lifetime lung cancer risk (1980 vital statistics)
= 4.584×10^{-4} per person

POP = 1980 US population
= 226,545,805

The final product of these values, propagating errors under the assumption that the distributions are log normal, is the range of 4,000 to 30,000.

The lifetime cancer mortality, M, associated with the presence of radon in drinking water, assuming that the risk is from inhaling radon, is calculated as (Crawford-Brown and Cothern 1987):

$$M = a_1 a_2 y_1 y_2 y_3 y_4$$

where

a_1 = the number of years of exposure to each member of the population using such water = 70 yr

a_2 = the number of working levels corresponding to a concentration of 100 pCi/L of air for Rn-222 in secular equilibrium with its short lived progeny = 1 WL/100 pCi/L (of air)

y_1 = the Rn-222 equilibrium factor = 0.5

y_2 = the equivalent occupational working months (number of 170 hr periods) per year for a member of the population = 18. This value has been reduced by a factor of 3 because we spend only a fraction of our time in homes and the breathing rate of the average person is less than that of a miner and there are uncertainties in the amount of unattached fraction.

y_3 = transfer factor for radon from water to air = 1×10^{-4}.

y_4 = the lifetime lung cancer mortality risk for exposure to 1 WLM, averaged over all ages of exposure.

Using the above method, the 95% confidence level range of estimated lung cancers due to radon in drinking water, with an average in the US of 200 to 600 pCi/L, is in the range of 80 to 800 per year, assuming that the errors propagate as for log normal distributions.

It must be concluded that there is a linear relationship. For example, Thomas et al. (1965) estimate curvilinearity to be raising the dose to the power 0.92 ± 0.07. The linear extrapolation is unlikely to underestimate the excess risk at low doses by more than a factor of 1.5 (Thomas et al. 1985).

C. Discussion

One question that generates discussion among those who study miners for effects of radon progeny is whether the model using relative risks or that using absolute risks fits the data better. The opinion has changed over the years, although at this time it appears to be the general consensus that neither or both fit the data. In any case, time of exposure needs to be included in the model, except for those prejudiced by the atomic bomb survivor data who prefer a relative risk model that is constant in time. The timing is important, because more cases appear early and for the old age group, and thus it is most detectable at both ends of the curve. No matter which model is used, one can get within a factor or two of a fit between the model and the data. Most feel that the relative risk model is easier to use and is likely more conservative.

If smoking and radon exposure interact additively or multiplicatively, then the relative or absolute model is the correct one. The problem is how to include in this model the decrease at higher ages, since that is the time when lung cancer is increasing in the general public but decreasing in miners since their exposure has stopped. The animal data appear to fit the relative risk model better.

If lung cancers are plotted as a function of time until 75 yr of age, it is proportional to the cube of age. A similar relationship appears to hold for asbestos exposure and mesothelioma. Such relationships suggest the validity of an absolute model. There is also an incentive to use the absolute model, if the regulatory requirement is expressed in the number of cases per 1000 in a lifetime.

Some complications exist in determining which is the best model in addition to those discussed above. The predictability of the model depends on the estimated latent period. The minimum latent period seems to be around 5 yr, but variation in this can have a large effect on the model prediction. Also a problem is such realities as miners stopping smoking because of the fear of lung cancer.

An often asked question is that if deaths are due to radon exposure to residences, then why did they not show up before the increase in lung cancers due to cigarette smoking in this century? The potential contribution to US lung cancer deaths between 1930 and 1987 has been examined using a relative risk model (Puskin and Yang 1988). The rate of radon induced lung mortality has been increasing sharply, however, the proportion attributable has remained fairly constant. It is estimated that 8–25% of the current lung cancer deaths are attributable to past radon exposure or roughly 8,000 to 26,000 using a relative risk coefficient range of 0.6 to 2.4% per WLM.

The estimates of risk coefficients in the range of exposure levels found in residences and the overall population risk estimates all have some uncertainty in them, because they are estimates based on projections or extrapolations of dose–response data into the unknown. In general the existing epidemiology data spans four orders of magnitude of dose, and the extrapolation is about one order of magnitude into the unknown. There is a general consensus that for environmental exposure levels found in residences, the uncertainty range is of the order of a factor of two to five. It is likely that this is an overestimate in the early years of life and an underestimate of risk in later life. Many feel that the main issue here is the effect of smoking. Are some of the deaths attributed to smoking really due to radon? This group would argue that studies therefore need to focus on the smokers. The opposing group would argue that non-smokers may not be contracting lung cancer from radon exposure, and thus the studies in the future should focus on the non-smokers. A possible contribution to the uncertainty can come from the well known healthy worker effect—that those in mines are healthier than the general public, and that they have better defenses against lung cancer in the 20–40 yr age range.

Although some of the uncertainty in risk estimates is due to statistical variation, the largest contribution is due to variations in exposure data and the lack of smoking histories.

Three internationally known and prestigious organizations have developed risk assessment models to estimate the risk due to exposure to radon (see Section VIIa): the National Academy of Sciences (the BEIR IV report from the Committee on Biological Effects of Ionizing Radiation), the International Commission for Radiation Protection (the ICRP 50 report) and the National Council on Radiation Protection and Measurements (the NCRP 78 report). It is fascinating to see such differences derived from much the same data. An important difference is that the BEIR IV analysis used the raw data while others used the published data. Most involved felt that these models agree within a factor of 2 which is as good as the data. The choice is, in some respects, one of the aesthetics and not science. If one prefers the relative risk model, the BEIR IV report is the most recent, includes information on the variances and has the actual data sets for the information used. Any model must include the effect of smoking and at least be consistent with the animal data. If the reader believes in threshold and repair, he chose the NCRP model. In extrapolating miner data to residential exposure, it should be noted that childhood exposure has no meaning and is a "lost dose."

The official EPA estimate of population risk due to the exposure to radon in residences in the US is in the range of 5,000 to 20,000/yr. Some argue that the upper limit is an overestimate, while those believing in thresholds would argue that the lower limit is zero. Thus if the lung really has a good capacity for cell repair, the range might be estimated to be 0–5,000 or 0–20,000 per year. However, this review estimates the population risk to be in the range of 4,000 to 30,000 annual lung cancer fatalities due to radon in indoor air.

VII. Risk Communication

As seen in the preceding sections, the assessment of risk due to indoor radon is generally agreed upon among the scientific community. That is, there is general agreement that exposure to radon, above the lowest levels involved in the epidemiology studies, causes lung cancer. And although there is some controversy about extrapolating the dose–response curve to residential levels, no one would argue that at least some residences have radon levels that produce a risk. The problem then is to communicate this to the public. This is a new area of investigation and few studies have been done to determine how best to communicate the information. Perhaps the biggest impediment is the fear many have concerning the effects of ionizing radiation. The public's perception of the health effects is clouded by human events such as the effects of nuclear weapons, and the events resulting from nuclear reactor accidents at Three Mile Island in the US, and Chernobyl in the USSR. There exists a kind of schizophrenic reaction, in that natural radiation from natural background, such as radon, and from sources such as drinking water, are somewhat acceptable; while exposure to X-rays receives a mixed reaction, and the lower level exposures to nuclear reactors and industrial activities are thought to be unacceptable.

Some of the problem in communication is related to the character of the radiation. To much of the public natural radiation exposure is more acceptable than man-made radiation (Litai 1980). Other characteristics of exposure also influence behavior. Is the exposure ordinary or catastrophic, voluntary or involuntary, delayed or immediate, controlled or uncontrolled, or regular or occasional? Although from the data available, the difference in acceptability between these categories seems to be an order of magnitude or so, the public tends to concern itself little with quantity and rejects exposures that are involuntary, man-made, catastrophic and immediate. Some perspective on the factors that cause cancers in the US can be found in the fact that tobacco, alcohol and diet account for almost 70% of them, while radon accounts for about 3% and other exposures to ionizing radiation much less (Upton 1988). Nevertheless, the public weighs causes such as toxic waste sites and other much smaller causes as more important (USEPA 1987c).

Some studies have been conducted on the effects of communication of the health effects of residential levels of radon (New Jersey 1986; Johnson and Luken 1987; USEPA 1987d; Smith and Johnson 1988; Johnson et al. 1988). These studies involve publics in Maine, New Jersey and New York. Space precludes a full discussion of the results of these studies, however, some of the general results are discussed. Perhaps the most important conclusion is that the general reaction of the public is apathy. More specifically households in Maine after receiving radon test results tended to understate the risks by orders of magnitude and showed no significant relationship between mitigating behavior and objective results. In New Jersey, those studied were unrealistically optimistic about radon and seriously underestimated the potential health effects. Despite widespread news stories, few people monitored their homes. The study in New York revealed that less than 1% of the homeowners contacted a government agency or public official after receiving radon test results. Other studies indicate that it does not really matter how the information is presented.

In a recent book examining the effect of communicating risk as a social process in several areas including radon, it was found that it was not enough for the scientific community to be aware nor was it sufficient for the issue to be validated in the social context (Krimszky and Plough 1988). It was observed that in the case of radon, what that little public reaction occurred resulted from active citizens in the affected areas and environmental activists pushing the problem into the national spotlight. They further observed that scientific uncertainty about radon risks has led to the dissemination of conflicting and ambiguous information. Since the problem is an individual one, the authors advocated educational programs to protect public health.

The area of risk communication is in its infancy, however, more work in this area is in progress. Clearly, if public health is to be best protected, more needs to be known about this important area.

Summary

This review concerns primarily the health effects that result from indoor air exposure to radon gas and its progeny. Radon enters homes mainly from the soil through cracks in the foundation and other holes to the geologic deposits beneath these structures. Once inside the home the gas decays (half-life 3.8 d) and the ionized atoms adsorb to dust particles and are inhaled. These particles lodge in the lung and can cause lung cancer. The introduction to this review gives some background properties of radon and its progeny that are important to understanding this public health problem as well as a discussion of the units used to describe its concentrations.

The data describing the health effects of inhaled radon and its progeny come both from epidemiological and animal studies. The estimates of risk from these two data bases are consistent within a factor of two. The epidemiological studies are primarily for hard rock miners, although some data exist for environmental exposures. The most complete studies are those of the US, Canadian, and Czechoslovakian uranium miners. Although all studies have some deficiencies, those of major importance include uranium miners in Saskatchewan, Canada, Swedish iron miners, and Newfoundland fluorspar miners. These six studies provide varying degrees of detail in the form of dose–response curves. Other epidemiological studies that do not provide quantitative dose–response information, but are useful in describing the health effects, include coal, iron ore and tin miners in the UK, iron ore miners in the Grangesburg and Kiruna, Sweden, metal miners in the US, Navajo uranium miners in the US, Norwegian niobian and magnitite miners, South African gold and uranium miners, French uranium miners, zinc-lead miners in Sweden and a variety of small studies of environmental exposure.

An analysis of the epidemiological studies reveals a variety of interpretation problem areas. The major and almost universal problem is in estimating exposure levels. In many cases there were no direct measurements of radon or radon progeny and the exposure levels are estimates based on irregular measurements and known levels in nearby mines.

Perhaps the most important variable or complicating factor in the determination of the risk due to radon exposure is the confounding factor of exposure to cigarette smoke. The general scientific concensus is that, although the interaction could be somewhere between linear and supramultiplicative, it is likely a combination, and closer to multiplicative.

A number of other complexities contribute to the uncertainty in the risk estimates, likely to a lesser degree than those of exposure measurements and cigarette smoke confounding. Mines have other contaminants in their air including asbestos, arsenic, chromium and diesel fumes. In some cases miners were exposed previous to their employment in the mines studied. The miner data are for adult males and it is not clear if there are differences for women and children. Many of the cohorts are still young and the miners have not been followed for a sufficient time. There is some suggestion that at lower doses the risk rate increases. It is known that repair mechanisms exists in the lung although their importance is not clear. There is the possibility that a threshold exists, but no supporting, statistically significant evidence yet exists. In some cases histopathological information exists, and it appears that there is a mix of cell types which is similar to that seen in lung cancers resulting from exposure to cigarette smoke.

Roughly 100,000 Americans die of lung cancer due to cigarette smoking each year and an additional 5,000 die from passive smoking. From the analysis in this review it is estimated that an additional 4,000 to 30,000 fatal lung cancers occur annually due to exposure to indoor radon and radon progeny.

Acknowledgment

The author wishes to thank Neal Nelson and Jim Smith for critical comments and discussions in the preparation of this review.

References

Abbatt JD, Newcombe HB (1981) Eldorado nuclear retrospective epidemiology project. A retrospective study of uranium workers from mines, mills, and refinery. Chap 58, pp 369-371, International Conference, Radiation Hazards in Mining: Control, Measurement, and Medical Aspects, Society of Mining Engineers of American Institute of Mining, Metallurgical, and Petroleum Engineers, Inc., New York, NY.

Altschuler B, Nelson N, Kuschner M (1964) Estimation of lung tissue dose from the inhalation of radon and daughters. Hlth Phys 10:1137-1161.

Archer VE, Wagoner JK, Lundin FE, Jr (1973a) Cancer mortality among uranium mill workers. J Occup Med 15:11-14.

Archer VE, Lundin FE, Wagoner JK (1973b) Lung cancer among uranium miners in the United States. Hlth Phys 25:351-371.

Archer VE (1984a) Enhancements of lung cancer by cigarette smoking in uranium and other miners. Proceedings of Symposium on Tumor Promotion and Enhancement in the Etiology of Human and Experimental Respiratory Tract Carcinogenesis, Marc Mass (ed). Raven Press, Boston.

Archer VE (1984b) Is silica or radon daughters the important factor in the excess lung cancer among underground miners? Proceedings of Symposium on Tumor Promotion and Enhancement in the Etiology of Human and Experimental Respiratory Tract Carcinogenesis, Marc Mass (ed). Raven Press, Boston.

Axelson O (1984) Room for a role for radon in lung cancer causation? Med Hypoth 13:51–61.

Axelson O, Sundell L (1978) Mining, lung cancer and smoking. Scand J Work Environ Hlth 4:46–52.

Bergman H, Axelson O (1983) Passive smoking and indoor radon daughter concentrations. Lancet 2:1308–1309.

Bergman H, Edling C, Axelson O (1986) Indoor radon daughter concentrations and passive smoking. Environ Int 12:17–19.

Berteig L, Stranden E (1981) Radon and radon daughters in mine atmospheres and influencing factors. Chap 17, pp 89–94. Gomez M (ed) International Conference on Radiation Hazards in Mining: Control, Measurement, and Medical Aspects, Society of Mining Engineers of American Institute of Mining, Metallurgical and Petroleum Engineers, Inc, New York, NY.

Boyd JT, Doll R, Faulds JS, Lieper J (1970) Cancer of the lung in iron ore (haematite) miners. Brit J Ind Med 27:97–105.

Crawford-Brown DJ, Cothern CR (1987) A Bayesian analysis or scientific judgement of uncertainties in estimating risk due to ^{222}Rn in US public drinking water supplies. Hlth Phys 53:4–21.

Chambers DB, Marchant RE (1985) Potential co-carcinogens in the uranium mine environment. Stocker H (ed) Proceedings of the International Conference on Occupational Radiation Safety in Mining, Canadian Nuclear Association, 111 Elizabeth Street, Toronto, Ontario, Canada, M5G 1P7, pp 615–622.

Chamaud J, Masse R, Morin M, Lafuma J (1985) Lung cancer induction by radon daughters in rats. Stocker H (ed) Proceedings of the International Conference on Occupational Radiation Safety in Mining, Canadian Nuclear Association, 111 Elizabeth Street, Toronto, Ontario, Canada, M5G 1P7, pp 350–353.

Clemente GF, Renzetti A, Santori G (1984) Relationship between the ^{210}Pb content of teeth and exposure to Rn and Rn daughters. Hlth Phys 47:253–262.

Cohen BL (1982) Failures and critique of the BEIR III lung cancer risk estimates. Hlth Phys 42:267–284.

Cohen BL (1984) Reply to Drs. Svec et al. Hlth Phys 46:964–965.

Cohen BL, Nelson D (1987) Radon, A Homeowner's Guide to Detection and Control. Consumers Union, Mount Vernon, NY.

Cohen BS (1987) Deposition of ultrafine particles in the human tracheobronchial tree. Pp 475–486. *In*: Radon and Its Decay Products: Occurrence, Properties and Health Effects, Hopke PK (ed) ACS Symp Ser 331, American Chemical Society, Washington, D.C.

Cote P, Townsend MG (1981) Mixtures of radon and thoron daughters in underground atmospheres. Hlth Phys 40:5–17.

Cothern CR, Crawford-Brown DJ, Wrenn ME (1988) Application of environmental dose-response models to epidemiology and animal data for the effects of ionizing radiation, submitted.

Cothern CR (1987) Estimating the health risk of radon in drinking water. J Am Waterworks Assoc, April:153–158.

Cothern CR, Smith JE Jr. (ed) (1987) Environmental Radon. Plenum Press, New York.

Cox DR (1972) Regression models and life-tables. J Royal Stat Soc, Series B 34, 187–202.

Crawford-Brown DJ, Cothern CR (1987) A bayesian analysis or scientific judgment of uncertainties in estimating risk due to ^{222}Rn in U.S. public drinking water supplies. Hlth Phys 53:11–21.

Crawford-Brown DJ (1988) The biokinetics and dosimetry of Rn-222 in the human body following ingestion of groundwater. Environ Geochem Hlth 11:10–17.

Cross FT (1988) A Radon Health Effects Literature Review: A Report to the US Department of Energy by Pacific Northwest Laboratory, Richland, WA 99352.

Damber L, Larsson LG (1982) Combined effects of mining and smoking in the causation of lung carcinoma. Acta Radiol Oncol 21:305–313.

Damber LA, Larsson LG (1987) Lung cancer in males and type of dwelling, An epidemiologic pilot study. Acta Oncol 26:211–215.

Dixon DW, James AC, Strong JC, Wrixon AD (1985) A review of all sources of exposure to natural radiation in UK mines. Stocker H (ed) Proceedings of the International Conference on Occupational Radiation Safety in Mining, Canadian Nuclear Association, 111 Elizabeth Street, Toronto, Ontario, Canada, M5G 1P7, pp 241–247.

Dungey CJ (1981) The radon problem of two Cornish tin mines. Chap 13, pp 65–68. Gomez M (ed) International Conference on Radiation Hazards in Mining: Control, Measurement, and Medical Aspects, Society of Mining Engineers of American Institute of Mining, Metallurgical and Petroleum Engineers, Inc, New York, NY.

Edling C (1982) Lung cancer and smoking in a group of iron ore miners. Am J Ind Med 3:191–199.

Edling C, Axelson O (1983) Quantitative aspects of radon daughter exposure and lung cancer in underground miners. Brit J Ind Med 40:182–187.

Federal Register (1986) Water pollution control: National Primary Drinking Water Regulations: advanced notice of proposed rulemaking, Tues, Sept 30, pp 34836–34862.

Fox AJ, Goldblatt P, Kinlen LJ (1981) A study of the mortality of Cornish tin miners. Brit J Ind Med 38:378–380.

George AC, Hinchliffe L (1972) Measurements of uncombined radon daughters in uranium mines. Hlth Phys 23:791–803.

George AC (1975a) Indoor and outdoor measurements of natural radon daughter decay products in New York City air, pp 741–750. *In:* The Natural Radiation Environment, II, CONF-720805, Adams JAS, Lowder WM and Gesell TF (eds) US Energy Research and Development Administration, Washington.

George AC, Hinchcliffe L, Sladowski R (1975b) Size distribution of radon daughter particles in uranium mine atmospheres. Am Ind Hyg Assoc J 36:484–490.

Ginevan ME, Mills WA (1986) Assessing the risks of Rn exposure: the influence of cigarette smoking. Hlth Phys 51:163–174.

Gleick J (1987) Chaos, Making a New Science. Viking Penguin, Inc, New York.

Guimond RJ, Ellett WH, Fitzgerald JE, Windham ST, Cury PA (1979) Indoor Radiation Exposure Due to Radium-226 in Florida Phosphate Lands, USEPA report EPA 520/4-78-013, Office of Radiation Programs, US Environmental Protection Agency, Washington, D.C., 20460.

Hallenbeck WH (1987) Risk assessment of exposure to airborne radon. Proceedings of the Air Pollution Control and Hazardous Work Management Meeting, New York, NY, June 21–26, Vol 87:41,3, pp 2–14.

Harley NH, Altman SM, Pasternack BS (1982) Genotoxic properties of radon and its daughters. Environ Sci Res 25:411–431.

Harley NH, Pasternak BS (1981) A model for predicting lung cancer risks induced by environmental levels of radon daughters. Hlth Phys 40:307-316.

Harley NH (1984a) Radon and lung cancer in mines and homes. New Eng J Med 310:1525-1527.

Harley NH (1984b) Comparing radon daughter dose: environmental versus underground exposure. Rad Prot Dos 7:371-375.

Harley NH, BS (1987) Updating radon daughter bronchial dosimetry. pp 419-429. *In*: Radon and Its Decay Products: Occurrence, Properties and Health Effects. Hopke PK (ed) ACS Symposium Series 331, American Chemical Society, Washington, D.C.

Harley NH (1988) Radon risk projection: validating the NCRP and other models. Proceedings of the 24th annual meeting of the National Council on Radiation Protection and Measurements, *in press*, 7910 Woodmont Avenue, Bethesda, MD 20814.

Hilding AC, Hilding D (1948) The volume of the bronchial tree at various levels and its possible physiologic significance. Ann Ontol Ther Laryng St Louis 65:324-342.

Hilding AC (1957) Ciliary streaming in the bronchial tree and the time element in carcinogenesis. New Eng J Med 256:634-640.

Hilding AC (1963) Phagocytosis, Mucous Flow and Ciliary Action. Arch Environ Hlth 6:61-73.

Horacek J, Placek V, Svec J (1977) Histologic types of bronchenic cancer in relation to different conditions of radiation exposure. Cancer 40:832-835.

Hornung RW, Samuels S (1981) Survivorship models for lung cancer mortality in uranium miners – is cumulative dose an appropriate measure of exposure? pp 363-368, Gomez M (ed) International Conference on Radiation Hazards in Mining: Control, Measurement, and Medical Aspects, Society of Mining Engineers of American Institute of Mining, Metallurgical and Petroleum Engineers, Inc, New York, NY.

Hornung RW, Meinhardt TJ (1987) Quantitative risk assessment of lung cancer in U.S. uranium miners. Hlth Phys 52:417-430.

Howe GR, Nair RC, Newcombe HB, Miller AB, Abbatt JD (1986) Lung cancer mortality (1950-1980) in relation to radon daughter exposure of a cohort of workers at the Eldorado Beaverlodge uranium mine. J Natl Cancer Inst 77:357-362.

Howe GR, Nair RC, Newcombe HB, Miller AB, Burch JD, Abbatt JD (1987) Lung cancer mortality (1950-80) in relation to radon daughter exposure in a cohort of workers at the Eldorado Port Radium uranium mine: possible modification of risk by exposure rate, J Natl Cancer Inst 79:1255-1260.

IARC (1988) IARC Monographs on the Evaluation of Carcinogenic Risk to Humans, Man-Made Mineral Fibers and Radon, Vol 43, IARC, Lyon, France.

ICRP (1981) International Commission on Radiological Protection, Limits for inhalation of radon daughters by workers, ICRP publ 32, Pergamon Press, New York.

ICRP (1987) International Commission on Radiological Protection, Lung cancer risk from indoor exposures to radon daughters, ICRP publ 50, Pergamon Press, New York.

International Atomic Energy Agency (1973) Inhalation Risks from Radioactive Contaminants, Tech Rep Ser, No. 142, International Atomic Energy Agency, Vienna.

Jablon S (1984) Epidemiologic perspectives in radiation carcinogenesis. *In*: Radiation Carcinogenesis: Epidemiology and Biological Significance, pp 1-19, Boice JD, Jr (ed) and Fraumini JF, Jr., Raven Press, New York.

Jablon S (1988) How to be Quantitative About Radiation Risk Assessments, Lecture 11, Laurison S. Taylor Lectures in Radiation Protection and Measurements, National Council on Radiation Protection and Measurments, 7910 Woodmont Ave, Bethesda, MD, 20814.

Jacobi W (1981) The new ICRP recommendations on occupational limits for radon daughters. Chap 78, pp 503–509, Gomez M (ed) International Conference on Radiation Hazards in Mining: Control, Measurement, and Medical Aspects, Society of Mining Engineers of American Institute of Mining, Metallurgical and Petroleum Engineers, Inc, New York, NY.

Jacobi W, Paretzke HG, Schindel F (1985) Lung cancer risk assessment of radon-exposed miners on the basis of a proportional hazard model. Stocker H (ed) Proceedings of the International Conference on Occupational Radiation Safety in Mining, Canadian Nuclear Association, 111 Elizabeth Street, Toronto, Ontario, Canada, M5G 1P7, pp 17–24.

James JC, Jacobi W, Steinhausler F (1981) Respiratory tract dosimetry of radon and thoron daughters: The state-of-the-art and implications for epidemiology and radiobiology. Chap 11, pp 42–53, Gomez M (ed) International Conference on Radiation Hazards in Mining: Control, Measurement, and Medical Aspects, Society of Mining Engineers of American Institute of Mining, Metallurgical and Petroleum Engineers, Inc, New York, NY.

James AC (1984) Dosimetric approaches to risk assessment for indoor exposure to radon daughters. Rad Prot Dos 7:353–366.

James AC (1987) A reconsideration of cells at risk and other key factors in radon daughter dosimetry, pp 400–418. *In*: Radon and Its Decay Products: Occurrence, Properties and Health Effects. Hopke PK (ed) ACS Symposium Series 331, American Chemical Society, Washington, D.C.

James AC (1988) Lung dosimetry. *In*: Radon and Its Decay Products in Indoor Air. Nazaroff WW and Nero AV, Jr (eds) John Wiley and Sons, New York.

Johnson FR, Luken RA (1987) Radon risk information and voluntary protection: evidence from a natural experiment. Risk Anal 7:97–107.

Johnson FR, Fisher A, Smith VK, Desvousges WH (1988) Informed choice or regulated risk? Lessons from a study in radon risk communication. Environment 30:12–35.

Johnson FR, Fisher A (in press) Conventional wisdom on risk communication and evidence from a field experiment. to be published in Risk Analysis.

Johnson JR, Leach VA (1981) An examination of the relationship between WLM exposure and dose. Chap 62, pp 390–397, Gomez M (ed) International Conference on Radiation Hazards in Mining: Control, Measurement, and Medical Aspects, Society of Mining Engineers of American Institute of Mining, Metallurgical and Petroleum Engineers, Inc, New York, NY.

Jorgensen HS (1973) A study of mortality from lung cancer among (iron ore) miners in Kituna 1950–1970. Work Env Hlth 10:126–133.

Krimsky S, Plough A (1988) Environmental Hazards: Communicating Risks as a Social Process. Auburn House Publishing Company, Dover, MA.

Kunz E, Svec J, Placek V (1978) Lung cancer mortality in uranium miners (methodological aspects). Hlth Phys 35:579–580.

Kunz E, Svec J, Placek V, Horacek J (1979) Lung cancer in man in relation to different time distribution of radiation exposure. Hlth Phys 36:699–706.

Land CE (1988) The ICRP 50 model. Proceedings of the 24th Annual Meeting of the National Council for Radiation Protection and Measurements, 7910 Woodmont Avenue, Bethesda, MD, 20814.

Lees REM, Steele R, Roberts JH (1987) A case-control study of lung cancer relative to domestic radon exposure. Int J Epidemiol 16:7-12.

Leira HL, Lund E, Refseth T (1986) Mortality and cancer incidence in a small cohort of miners exposed to low levels of alpha radiation. Hlth Phys 50:189-194.

Lerchen ML, Wiggins CL, Samet JM (1987) Lung cancer and occupation in New Mexico. J Natl Cancer Inst 79:639-645.

Litai D (1980) A Risk Comparison Methodology for the Assessment of Acceptable Risk, PhD Thesis, Massachusetts Institute of Technology.

Lubin JH (1988) On the BEIR IV lung cancer risk projection model for radon exposure. Proceedings of the 24th Annual Meeting of the National Council for Radiation Protection and Measurements, 7910 Woodmont Avenue, Bethesda, MD, 20814.

Lubin JH (1988a) Models for the analysis of radon exposed populations. Yale J Biol Med 61:195-214.

Lundin FE Jr., Lloyd JW, Smith EM, Archer VE, Holaday DA (1969) Mortality of uranium miners in relation to radiation exposure, hard-rock mining and cigarette smoking −1950 through 1967. Hlth Phys 16:571-578.

Lundin FE, Wagoner JK, Archer VE (1971) Radon daughter exposure and respiratory cancer. *In*: Quantitative and Temporal Aspects. United States Department of Health, Education and Welfare (USDHEW), National Institute for Occupational Safety and Health (NIOSH), National Institute of Environmental Health Services (NIEHS), Joint Monograph, No. 1.

Martel EA (1975) Tobacco radioactivity and cancer in smokers. Am Sci 63:404-412.

Martel EA (1982a) The natural alpha radiation environment: a preliminary assessment, pp 121-130, Proceedings of the Natural Radiation Environment, Vohra KG (ed). Wiley Eastern Ltd., New Delhi.

Martel EA, Poet SE (1982b) Radon progeny on biological surfaces and their effects, pp 283-289. *In*: Proceedings of the Natural Radiation Environment, Vorha KG (ed). Wiley Eastern Ltd., New Delhi.

Martel EA (1983) Alpha-radiation dose at bronchial bifurcations of smokers from indoor exposure to radon progeny. Proc Natl Acad Sci USA 80:1285-1289.

Martel EA (1985) Enhanced ion production in convective storms by transpired radon isotopes and their decay products. J Geophys Res 90:5909-5916.

Martel EA (1987) Critique of current dosimetry models for radon progeny exposure, pp 444-461. *In*: Radon and Its Decay Products: Occurrence, Properties and Health Effects. Hopke PK (ed) ACS Symposium Series 331, American Chemical Society, Washington, D.C.

Martel EA (submitted) Spontaneous mutations attributable to inhaled and ingested radioisotopes. Submitted to Ambio.

Martonen TB, Hofmann W, Lowe JE (1987) Cigarette smoke and lung cancer. Hlth Phys 52:213-217.

McCracken WJ (1981) Radon gas, bronchogenic carcinoma−Ontario experience, Chap 122, pp 819-822. Gomez M (ed) International Conference on Radiation Hazards in Mining: Control, Measurement, and Medical Aspects. Society of Mining Engineers of American Institute of Mining, Metallurgical and Petroleum Engineers, Inc, New York, NY.

Morrison HI, Wigle DT, Stocker H, deVilliers AJ (1981) Lung cancer mortality and radiation exposure among the Newfoundland fluorspar miners, pp 372–376, Gomez M (ed). *In*: Radiation Hazards in Mining. American Institute of Mining, Metallurgical and Petroleum Engineers Inc, New York.

Morrison HI, Semenciw RM, Mao Y, Corkill DA, Dory AB, deVillers AJ, Stocker H, Wigle DT (1985) Lung cancer mortality and radiation exposure among the Newfoundland fluorspar miners, pp 365–368. *In*: Occupational Radiation Safety in Mining. Stocker H (ed) Canadian Nuclear Association, 111 Elizabeth Street, 11th Floor, Toronto, Ontario, Canada M5G 1P7.

Muller C, Ruzicka L, Bakstein J (1967) The sex ratio in the offspring of uranium miners. Univ Carolinae Med 13:599–603.

Muller J, Wheeler WC, Gentleman JF, Suranyi G, Kusiak R, Smith M (1981) The Ontario miners – mortality study. Chap 56, pp 359–362, Gomez M (ed). *In*: Radiation Hazards in Mining. American Institute of Mining, Metallurgical and Petroleum Engineers Inc, New York.

Muller J, Wheeler WC, Gentleman JF, Suranyi G, Kusiak RA (1983) Study of mortality of Ontario miners 1955–1977. Ontario Ministry of Labour, Ontario Workers' Compensation Board, Atomic Energy Control Board of Canada, Toronto, Ontario M7A 1T7.

Muller J, Wheeler WC, Gentleman JF, Suranyi G, Kusiak RA (1985) Study of Mortality of Ontario Miners. *In*: Stocker H (ed) Proceedings of the International Conference on Occupational Radiation Safety in Mining. Canadian Nuclear Association, 111 Elizabeth Street, Toronto, Ontario, Canada M5G 1P7, pp 335–343.

Myers DK, Johnson JR (1985) Impact of occupational hazards on the life expectancy of uranium miners. *In*: Stocker H (ed) Proceedings of the International Conference on Occupational Radiation Safety in Mining. Canadian Nuclear Association, 111 Elizabeth Street, Toronto, Ontario, Canada M5G 1P7, pp 627–634.

Nair RC, Abbatt JD, Howe GR, Newcombe HB, Frost SE (1985) Mortality experience among workers in the uranium industry. *In*: Stocker H (ed) Proceedings of the International Conference on Occupational Radiation Safety in Mining. Canadian Nuclear Association, 111 Elizabeth Street, Toronto, Ontario, Canada M5G 1P7, pp 354–364.

NAS (1980) National Academy of Sciences, The effects on populations of exposure to low levels of ionizing radiation (BEIR III). National Academy Press, 2101 Constitution Ave., N.W., Washington, D.C. 20418.

NAS (1988) National Academy of Sciences, Health risks of radon and other internally deposited alpha-emitters (BEIR IV). National Academy Press, 2101 Constitution Ave., N.W., Washington, D.C. 20418.

NCRP (1984a) National Council on Radiation Protection and Measurements, Exposures from the uranium series with emphasis on radon and its daughters. NCRP rept 77, 7910 Woodmont Avenue, Bethesda, MD 20814.

NCRP (1984b) National Council on Radiation Protection and Measurements. Evaluation of occupational and environmental exposures to radon and radon daughters in the United States. NCRP rept 78, 7910 Woodmont Avenue, Bethesda, MD 20814.

NCRP (1987a) National Council on Radiation Protection and Measurements. Ionizing Radiation Exposure of the Population of the United States. Rept 93, 7910 Woodmont Avenue, Bethesda, MD 20814.

NCRP (1987b) National Council on Radiation Protection and Measurements. Exposure of the Population in the United States and Canada from Natural Background Radiation. Rept 94, 7910 Woodmont Avenue, Bethesda, MD 20814.

Nero AV, Schwehr MB, Nazaroff WW, Revzan KL (1986) Distribution of airborne radon-222 concentrations in US homes. Science 234:992–997.

New Jersey (1987) Public Response to the Risk from Radon, 1986. New Jersey Department of Environmental Protection, Division of Environmental Quality.

NIOSH (1987) National Institute for Occupational Safety and Health, U.S. Department of Health and Human Services. Radon Progeny in Underground Mines, Criteria for a Recommended Standard – Occupational Exposure to Radon Progeny in Underground Mines. Public Health Service, Centers for Disease Control, National Institute for Occupational Safety and Health, Division of Standards Development and Technology Transfer, available from the Superintendent of Documents, U.S. Government Printing Office, Washington, D.C., 20402.

O'Riordan MC, Rae S, Thomas GH (1981) Radon in British mines – a review. Chap 15, pp 74–81. *In*: Radiation Hazards in Mining, Gomez M (ed), American Institute of Mining, Metallurgical and Petroleum Engineers Inc, New York.

Paschoa AS, Nobrega AW (1981) Non-nuclear mining with radiological implications in Araxa. Chap 16, pp 82–88. *In*: Radiation Hazards in Mining, Gomez M (ed), American Institute of Mining, Metallurgical and Petroleum Engineers Inc, New York.

Phillips CR, Khan A, HM (1988) The nature and determination of the unattached fraction of radon and thoron progeny. *In*: Radon and Its Decay Products in Indoor Air, Nazaroff WW and Nero AV Jr. (eds), John Wiley and Sons, New York.

Plough A, Krimsky S (1987) The emergence of risk communication studies: Social and political context. Sci Technol Human Values 12:4–10.

Pochin EE (1985) The epidemiology of radiation effects (1985). *In*: Proceedings of the International Conference on Occupational Radiation Safety in Mining, Stocker H (ed), Canadian Nuclear Association, 111 Elizabeth Street, Toronto, Ontario, Canada M5G 1P7, pp 167–172.

Pohl-Ruling J, Scheminsky (1972) The natural radiation environment of Badgestein, Austria and its biological effect. The Natural Radiation Environment, II. Adams JAS, Lowder WM (eds) US Department of Energy, CONF 720805P1.

Pohl E, Pohl-Ruling J, Steinhausler F (1977) The natural radioactivity of the air in the region of Badgastein/Austria, the measurement of its local and temporal fluctuations and the resulting dose to various population groups. International Symposium on Areas of High Natural Radioactivity. Edited by Academia Brasileria de Ciencias Rio de Janeiro, RJ, p 182.

Pohl-Ruling J, Fischer P (1983) Chromosome aberrations in inhabitants of areas with elevated natural radioactivity. *In*: Radiation-Induced Chromosome Damage in Man. pp 527–560. Alan R. Liss, Inc, 150 Fifth Avenue, New York, NY, 10011.

Pohl-Ruling J, Fischer P, Pohl E (1987) Effect of peripheral blood chromosomes. pp 487–501. *In*: Radon and Its Decay Products: Occurrence, Properties and Health Effects, Hopke PK (ed), ACS Symposium Series 331, American Chemical Society, Washington, D.C.

Puskin JS, Yang Y (1988) A retrospective look at Rn-induced lung cancer mortality from the viewpoint of a relative risk model. Hlth Phys 54:635–643.

Radford EP, Martel EA (1977) Polonium-210: Lead–210 ratios as an index of residence times of insoluble particles from cigarette smoke in bronchial epithelium. pp 567–580. *In*: Inhaled Particles IV, Walton WH (ed), Part 2, proceedings of an annual international symposium organized by the British Occupational Hygiene Society in Edinburgh, Scotland, 22–265 Sept 1975. Pergamon Press, Oxford.

Radford EP (1984) Radiogenic cancer in underground miners. *In*: Radiation Carcinogenesis: Epidemiology and Biological Significance, Boice JD Jr. and Fraumeni JF (eds), pp 225–230. Raven Press, New York.

Radford EP (1985) Potential health effects of indoor radon exposure. Environ Hlth Pers 62:281–287.

Radford EP, Renard EGSC (1984) Lung cancer in swedish iron miners exposed to low doses of radon daughters. New Eng J Med 310:1485–1494.

Repace JL, Lowrey AH (1985) A quantitative estimate of non-smokers' lung cancer risk from passive smoking. Environ Int 11:3–12.

Reubel B, Atmuller C, Steinhausler F, Huber W (1987) Biophysical effects of radon exposure on human lung cells. pp 502–512. *In*: Radon and Its Decay Products: Occurrence, Properties and Health Effects, Hopke PK (ed), ACS Symposium Series 331, American Chemical Society, Washington, D.C.

Rockette HE (1977) Cause specific mortality of coal miners. J Occup Med 19:795–801.

Rockette HE (1980) Mortality Patterns of Coal Miners. *In*: Health Implications of New Energy Technologies, Rom WM and Archer VE (eds), Ann Arbor Science Publishers Inc.

Saccamanno G, Huth GC, Auerbach O, Kuschner M (1988) Relationship of radioactive radon daughters and cigarette smoking in the genesis of lung cancer in uranium miners. Cancer (October 1).

Samet JM, Kutvirt DM, Waxweiler RJ, Key CR (1984) Uranium mining and lung cancer in Navajo men. New Eng J Med 310:1481–1484.

Samet JM, Morgan MV, Key CR (1985) Studies of uranium miners in New Mexico. *In*: Proceedings of the International Conference on Occupational Radiation Safety in Mining, Stocker H (ed), Canadian Nuclear Association, 111 Elizabeth Street, Toronto, Ontario, Canada M5G 1P7, pp 623–626.

Sciocchette G, Scacco F, Clemente GF (1981) The radiation hazards in Italian nonuranium mines – aspects of radiation protection. Chap 14, pp 69–73. *In*: Radiation Hazards in Mining, Gomez M (ed), American Institute of Mining, Metallurgical and Petroleum Engineers Inc, New York.

Sevcova M, Svec J, Thomas J (1978) Alpha irradiation of the skin and the possibility of late effects. Hlth Phys 35:803–806.

Solli HM, Anderson A, Strauden E, Langard S (1985) Cancer incidence among workers exposed to radon and thoron daughters at a niobium mine. Scand J Work Environ Hlth 11:7–13.

Smith VK, Johnson FR (1988) How do risk perceptions respond to information? The case of radon. Rev Econ Stat 52:1–8.

Smith ME (1981) Health hazards in mining – the files and facilities. Chap 125, pp 836–841. *In*: Radiation Hazards in Mining, Gomez M (ed), American Institute of Mining, Metallurgical and Petroleum Engineers Inc, New York.

Smith VK, Desvousges WH, Fisher A, Johnson FR (1987) Communicating radon risk effectively: a mid-course evaluation. EPA-230-07-87-029, Office of Policy Analysis, US Environmental Protection Analysis, Washington, D.D. 20460.

Steinhausler F, Pohl E (1973) The concentration of ^{222}Rn, ^{220}Rn and their daughters in the air, the dependence on meteorological variables and contribution to the radiation dose for the inhabitants of a radon spa. *In*: Health Physics Problems of Internal Contamination. Proceedings of the IRPA Second European Congress of Radiation Protection, Bujdoso E (ed), Akademiai Kiado, Budapest, pp 397–400.

Steinhausler F (1985) The radon dilemma. *In*: Proceedings of the International Conference on Occupational Radiation Safety in Mining, Stocker H (ed), Canadian Nuclear Association, 111 Elizabeth Street, Toronto, Ontario, Canada M5G 1P7, pp 637–641.

Steinhausler F, Hofman W (1985) Inherent dosimetric and epidemiological uncertainties association with lung cancer risk assessment for mining populations. *In*: Proceedings of the International Conference on Occupational Radiation Safety in Mining, Stocker H (ed), Canadian Nuclear Association, 111 Elizabeth Street, Toronto, Ontario, Canada M5G 1P7, pp 327–333.

Steinhausler F (1987) The validity of risk assessments for lung cancer induced by radon daughters. pp 430–443. *In*: Radon and Its Decay Products: Occurrence, Properties and Health Effects, Hopke PK (ed), ACS Symposium Series 331, American Chemical Society, Washington, D.C.

Steinhausler F (1988) Epidemiology evidence of radon-induced health risks. *In*: Radon and Its Decay Products in Indoor Air, Nazaroff WW and Nero AV Jr (eds), John Wiley and Sons, New York.

Stranden E (1985) Thoron daughter to radon daughter ratios in mines. *In*: Proceedings of the International Conference on Occupational Radiation Safety in Mining, Stocker H (ed), Canadian Nuclear Association, 111 Elizabeth Street, Toronto, Ontario, Canada M5G 1P7, pp 604–606.

Stranden E (1986) Radon in Norwegian dwellings and the feasibility of epidemiological studies. Rad Environ Biophys 25:37–42.

Subba MC, Vohra KG (1985) A study of the dose conversion factors for inhalation risk assessment from radon daughters in mine atmospheres. *In*: Proceedings of the International Conference on Occupational Radiation Safety in Mining, Stocker H (ed), Canadian Nuclear Association, 111 Elizabeth Street, Toronto, Ontario, Canada M5G 1P7, pp 131–133.

Svec J, Kunz E, Placek V (1976) Lung cancer in uranium miners and long-term exposure to radon daughter products. Hlth Phys 30:433–437.

Svec J, Kunz E, Placek V, Smid A (1984) Comments on lung cancer risk estimates. Hlth Phys 46:961–964.

Svec J, Kunz E, Tomasek L, Placek V, Horacek J (1988) Cancer in man after exposure to Rn daughters. Hlth Phys 54:27–46.

Svensson C, Eklund G, Pershagen G (1987) Indoor exposure to radon from the ground and bronchial cancer in women. Int Arch Occup Environ Hlth 59:123–131.

Tirmarche M, Brenot J, Piechowski J, Chameaud J, Pradel J (1985) The present state of an epidemiology study of uranium miners in France. pp 344–349. *In*: Proceedings of the International Conference, Vol 1, Occupational Radiation Safety in Mining, Stocker H (ed), Toronto Canadian Nuclear Association.

Thomas DC, McNeill KG (1982) Risk estimates for the health effects of alpha radiation. Atomic Energy Control Board, Ottawa, Canada.

Thomas DC, McNeill KG, Dougherty C (1985) Estimates of lifetime lung cancer risks resulting from Rn progeny exposure. Hlth Phys 49:825–846.

United Nations (1982) Ionizing radiation: sources and biological effects. United Nations committee on the effects of atomic radiation, United Nations, New York.

Upton AC (1988) Carcinogenic risk assessment in proper perspective. Toxicol Ind Hlth 4:443–452.

USEPA (1979) Indoor Radiation Exposure Due to Radium-226 In Florida Phosphate Lands. EPA 520/4-78/013, Office of Radiation Programs, Washington, D.C. 20460.

USEPA (1986) Federal Register 40 CFR 141 Water Pollution Control; National Primary Drinking Water Regulations; Radionuclides; Advanced Notice of Proposed Rulemaking. Tues, Sept 30, pp 34836–34862.

USEPA (1987a) Radon reference manual. EPA/520/1-87-20, Office of Radiation Programs, Washington, D.C. 20460.

USEPA (1987b) Summary of state radon programs. EPA/520/1-87-19-1, Office of Radiation Programs, Washington, D.C. 20460.

USEPA (1987c) Unfinished Business: A Comparative Assessment of Environmental Problems. Office of Policy Analysis, US Environmental Protection Agency, Washington, D.C. 20460.

USEPA (1988a) Radon reduction techniques for detached houses. Technical guidance (2nd Ed), EPA/625/5-87/019, Office of Research and Development, Washington, D.C. 20460.

USEPA (1988b) Indoor Radon Pollution, a Bibliography (the earlier version was EPA IMSD/86-002).

Uzunov I, Steinhausler F, Pohl E (1981) Carcinogenic risk of exposure to radon daughters associated with radon spas. Hlth Phys 41:807–813.

Valdivia AS (1981) Medical surveillance program for uranium workers in Grants, New Mexico. Chap 124, pp 831–835. *In*: Radiation Hazards in Mining, Gomez M (ed), American Institute of Mining, Metallurgical and Petroleum Engineers Inc, New York.

Wagoner JK, Miller RW, Lundin FE, Fraumeni JF, Haij ME (1963) Unusual cancer mortality among a group of underground metal miners. New Eng J Med 269:284–289.

Waxweiler RJ, Roscoe RJ, Archer VE, Thun MJ, Wagner JK, Lundin FE, Jr (1981) Mortality followup through 1977 of the white underground miners cohort examined by the U.S. Public Health Service. pp 823–830. *In*: Radiation Hazards in Mining, Gomez M (ed), American Institute of Mining, Metallurgical and Petroleum Engineers, Inc, New York.

Whittemore AS, McMillan A (1983) Lung cancer mortality among U.S. uranium miners: a reappraisal. J Natl Cancer Inst 71:489–499.

Wiese WH (1981a) Birth effects in areas of uranium mining. pp 605–607. *In*: Radiation Hazards in Mining, Gomez M (ed), American Institute of Mining, Metallurgical and Petroleum Engineers, Inc, New York.

Wiese WH (ed) (1981b) Birth Effects in the Four Corners Area. Transcript of a meeting, Department of Family, Community and Emergency Medicine, University of New Mexico School of Medicine, Albuquerque, NM.

Wiese WH, Skipper BJ (1986) Survey of reproductive outcomes in uranium and potash mine workers: results of first analysis. Ann Am Conf Gov Ind Hyg 14:187–192.

Wilkening MH (1977) Radon -222 concentrations in the Carlsbad Caverns. International Symposium on Areas of High Natural Radioactivity. Edited by Academia Brasileria de Ciencias, Rio de Janeiro, RJ, p 183.

World Health Organization (WHO) (1982) The World Health Organization histological typing of lung tumors. Am Soc Clin Pathol 2:123–136.

Manuscript received November 29, 1988; accepted February 25, 1989.

Ecological Toxicology and Human Health Effects of Heptachlor

E.A. Fendick*, E. Mather-Mihaich*, K.A. Houck*, M.B. St. Clair*, J.B. Faust*, C.H. Rockwell*, and M. Owens*

Contents

I. Introduction	61
II. Physical Properties and Synthesis	63
III. Sampling and Analysis	64
IV. Transport, Transformation, and Environmental Concentrations	67
V. Ecological Toxicology	74
A. Terrestrial Biota	74
B. Aquatic Animals	92
C. Bioaccumulation	96
VI. Physiological Properties of Heptachlor	98
A. Metabolism	98
B. Absorption, Distribution, and Elimination	100
VII. Organ System Toxicity	101
VIII. Mutagenicity	106
IX. Carcinogenicity	107
X. Teratogenicity	111
XI. Toxicology and Human Exposure	113
A. Acute Toxicity	113
B. Chronic Human Exposure	115
C. Epidemiology	116
XII. Recommendations	117
Summary	118
References	119

I. Introduction

The chlorinated cyclodiene heptachlor was granted US registration as an agricultural and domestic insecticide in 1952. Through the mid-1960s, it was primarily employed in soil applications against agricultural pests and termites (Gosselin et al. 1976; Murphy 1986). In 1962, Canada banned heptachlor as a seed treatment

---

*Integrated Case Studies in Toxicology 1987, Integrated Program in Toxicology, Duke University, Durham, North Carolina 27710.

Table 1. United States usage of heptachlor in various years

Use	kg × 10⁶ (% of total)				
	1971[a]	1972[b]	1973[c]	1974[c]	1975–76[a,d]
Corn	1.96 (68.1)		0.51 (57.3)	0.54 (58.1)	2.59 (57.9)
Seed dressing			0.09 (10.1)	0.12 (12.9)	0.59 (13.2)
PCO[e]			0.28 (31.5)	0.25 (26.9)	1.20 (26.8)
Miscellaneous[f]			0.01 (1.1)	0.02 (2.2)	0.09 (2.0)
Totals	2.88	6.0	0.89	0.93	4.47

[a] WHO 1984b.
[b] Lu et al. 1975.
[c] From Train 1976.
[d] July 1975–December 1976.
[e] Pesticide control operators.
[f] Includes use on pineapple (approx. 4500 kg annually) and several thousand kg used on citrus.

(McEwen and Stephenson 1979), while in Japan, the only accepted use for heptachlor was in termite control (IARC 1974). According to the FAO/WHO (1971), the relative worldwide use of heptachlor in 1970 was 5% in Africa, 5% in Canada and the US, 15% in Asia, 15% in South America, and 60% in Europe.

As an agricultural insecticide, heptachlor was applied to seeds and bulbs (to prevent insect damage prior to germination) as well as to hundreds of thousands of acres of pineapple, corn, and citrus. With agricultural consumption of about 550,000 kg annually in the early 1970s, heptachlor represented about 0.5% of the total US agrarian insecticide use. An additional 250,000 kg were applied in 1974 for industrial and commercial purposes (Train 1976). (See Table 1 for further uses.) Domestic uses of heptachlor included application to soils as a termiticide, to gardens and lawns for grub, chigger, and tick control, and as a component of shelf paper. Of these, it was most valuable in controlling mealy bug infestation of pineapple crops, and in termite control. As of August 1, 1976, the US Environmental Protection Agency (EPA) had cancelled heptachlor registration for all uses except subterranean termite control, seed and narcissus bulb treatment, and fire ant control primarily on pineapple crops (Train 1976). Heptachlor use was also restricted in Italy and Switzerland (IARC 1974), and the USSR (IRPTC 1982).

Heptachlor was considered essential for growing pineapple in Hawaii, since it, as well as mirex, are extremely effective in fire ant control. Fire ants spread viruses from the mealy bug to pineapple plants, resulting in lethal mealy bug wilt. Diazinon and malathion are marginally effective against the mealy bug itself, but ineffective in controlling the ants. The EPA permitted the continued application of heptachlor to pineapple, even though there is a zero tolerance level for heptachlor or heptachlor epoxide residues on fruit. The reasons were severalfold:

mealy bug wilt was under control using heptachlor, there were uncertainties about the continued use of mirex, existing chemical alternatives (e.g., pyrethrins, malathion, methoxychlor) were ineffective, and the quantity of heptachlor was believed to be too small to arouse great concern (Train 1976). Despite these arguments, registration of heptachlor for use on pineapple was cancelled as of December 31, 1982 (USPHS 1987).

By July 1983, US heptachlor application was limited to subterranean termite control and treatment of power and telephone pedestals for fire ant control (USPHS 1987). In 1985, heptachlor alone, or in combination with chlordane, accounted for 60-65% of the termiticides used in the US (EPA 1987). In 1987, the EPA and the Agency for Toxic Substances and Disease Registry classified heptachlor as a Priority Group 1 Hazardous Substance, making eligible the immediate dispensation of Superfund money in cleanup of heptachlor-contaminated sites. On August 11, 1987, Velsicol Chemical Corporation, the only US licensed manufacturer of chlordane and heptachlor, agreed to cease the sale of pesticides containing chlordane and heptachlor until new application methods could be developed which would not produce detectable airborne levels inside homes (EPA 1987).

As questions have arisen about possible adverse health effects resulting from exposure to heptachlor, a desire for effective alternative termiticides has arisen. The existing alternatives to heptachlor are, however, not considered as effective or as persistent; many of these compounds (e.g., aldrin/dieldrin, endrin/endrin aldehyde) also have health risks. With the above background, this review of the impacts of heptachlor on the environment, on laboratory animals, and man was initiated.

II. Physical Properties and Synthesis

Heptachlor has the molecular formula $C_{10}H_5Cl_7$ and a molecular weight of 373.35. The Chemical Abstracts name is 1,4,5,6,7,8,8,-heptachloro-3a,4,7,7a-tetrahydro-4,7- methano-1H-indene. Heptachlor is a white crystalline solid with a mild odor of camphor, a melting point of 93°C (46-74°C for the technical product), and a density of 1.65-1.67 g ml^{-1} at 25°C. It has a boiling point of 135-145°C and a vapor pressure of 4×10^{-4} mm Hg at 25°C (WHO 1984b).

Heptachlor is virtually insoluble in water (0.056 mg L^{-1}), but fairly soluble in organic solvents: ethanol (45 g L^{-1}), acetone (750 g L^{-1}), xylene (1020 g L^{-1}), and benzene (1060 g L^{-1}). It is stable to visible light, air, moisture, and moderate heat (160°C) (WHO 1984b). Exposure of heptachlor to ultraviolet light causes chemical rearrangement to various photoisomers (Ivie et al. 1972; Knox et al. 1973; McGuire et al. 1972; Parlar et al. 1978; Podowski et al. 1979), some of which exhibit greater insecticidal activity than heptachlor itself (Rosen et al. 1969).

The basic carbon skeleton of many of the chlorinated insecticides is synthesized by a Diels-Alder reaction in which an unsaturated molecule is added across

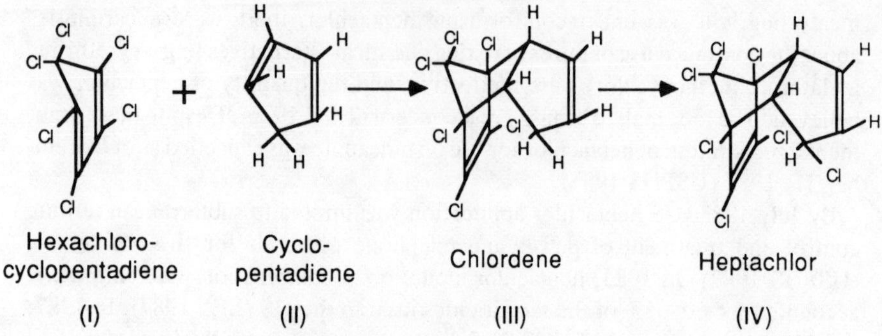

Fig. 1. Schematic summary of heptachlor synthesis.

Hexachloro-cyclopentadiene (I) + Cyclopentadiene (II) → Chlordene (III) → Heptachlor (IV)

a diene system. In the case of heptachlor and other cyclodiene insecticides, hexachlorocyclopentadiene [I] and cyclopentadiene [II] are reacted under relatively mild conditions to produce chlordene [III] as shown in Fig. 1. Chlorination of this adduct [III] with CCl_4 and Cl_2 at 70°C followed by distillation yields predominantly the octachloro-dihydro-cyclopentadiene derivatives such as chlordane, while chlorination of [III] utilizing CCl_4 with SO_2Cl_2 yields heptachlorodicyclopentadiene [IV] (heptachlor) (Heys et al. 1979; Hyman 1949, 1951).

Today, heptachlor is produced commercially via a similar chlorination of chlordene in the presence of a catalyst such as Fuller's earth at 0–5°C. The solvent is distilled off, and the residue recrystallized from methanol before grinding (WHO 1984b). Synthesis of both enantiomers of heptachlor has been accomplished through chlorination (Cl_2 in $CHCl_3$-CCl_4) of optically pure 1-hydroxychlordene. The corresponding optically pure epoxides were prepared via oxidation of the heptachlor enantiomers with CrO_3. The insecticidal activity of these enantiomers was measured on male adult German cockroaches, revealing that the racemic form of heptachlor exhibited a stronger activity than either of its enantiomers. The (+) enantiomer of heptachlor epoxide showed a stronger activity than its corresponding optical antipode but the difference was relatively small [2.3 X] (Miyazaki et al. 1978, 1980).

While careful synthesis and separation will yield relatively pure heptachlor or its octachloro-analog chlordane, the commercial products, technical heptachlor and chlordane, are mixtures of the pure compound plus related reaction products (Cochrane and Greenhalgh 1976; March 1952). A variety of commercial products was available through 1987 as dusts, granules, and emulsifiable concentrates of varying strengths. In general, these products contained heptachlor and related compounds in approximately a 5:2 ratio (Morgan 1987).

III. Sampling and Analysis

Methods for extraction and purification of heptachlor residues from various physical and biological sources are summarized in Table 2.

Table 2. Summary of extraction and purification methods for the determination of heptachlor residues in samples

Method	Reference
Heptachlor extraction	
Solvent	Alford-Stevens et al. 1986
	Ambrus et al. 1981a
	Barquet et al. 1981
	Brodie et al. 1984
	Frank et al. 1982
	Javadi and Hajari 1986
	McNeil et al. 1977
	Rihan et al. 1978
	Sherma and Shafik 1975
	Stanley et al. 1971
	Waliszewski and Szymczynski 1985
Macroreticular Resin	Burnham et al. 1972
	Livingstone and Jones 1981
	McNeil et al. 1977
	Strachan 1985
Polyurethane Foam	Gesser et al. 1971
	Wright and Leidy 1982
Carbon	Grob and Zurcher 1976
Purification procedures	
Florisil	Feroz and Khan 1979
	Solomon 1979
	Stanley et al. 1971
	Veith et al. 1981
Alumina	Javadi and Hajari 1986
	Zubillaga et al. 1987
Distillation	Murray 1979
	Nash 1984

Many different trapping methods extract heptachlor from air samples. Most rely on a solvent-charged impinger; however, newer solid phase samplers have proven quite reliable (Thomas and Seiber 1974). Hexylene glycol is the most common liquid trapping agent (Sherma and Shafik 1975; Stanley et al. 1971). Polyurethane foam plugs (Wright and Leidy 1978) and Chromosorb 102 (Livingstone and Jones 1981) are the two most common solid trapping devices. The detection limit using the polyurethane foam plus is 0.01 µg (Wright and Leidy 1978). In all cases, a pump pulls air at a measured rate through the trapping agent.

After the pesticide is trapped, an extraction agent removes the compound from the adsorbent. Commonly used solvents are acetone, acetonitrile, methanol, methylene chloride, or ethyl acetate (Arthur et al. 1976; Sherma and Shafik

1975; Stanley et al. 1971; Wright and Leidy 1978). The extract is concentrated and purified by column chromatography through florisil or silica gel (Sherma and Shafik 1975; Stanley and Post 1967). Elution is accomplished using either ethyl and petroleum ethers or methylene chloride and hexane with acetonitrile (Luke and Matsumoto 1986). This final extract is then ready for quantification.

Soil and sediment samples are dried to a constant weight by filtration, centrifugation, or air- or oven-drying (Ambrus et al. 1981b; Frank et al. 1981; Herzel 1972; Waliszewski and Szymczynski 1985). Samples are well mixed, usually in a blender, and extracted with acetone, hexane, acetone/light petroleum, or similar solvents (Townsend and Specht 1975). A Soxhlet apparatus is used routinely in this extraction step (Herzel 1972; Nash 1984; Zubillaga et al. 1987). Column chromatography purifies the concentrated sample; alumina and florisil are common packing materials, with hexane as the eluting liquid (Herzel 1972; Teichman et al. 1978; Zubillaga et al. 1987). Samples may be concentrated further by evaporation, and then quantified.

Water samples, collected in solvent-washed jars, are mixed with a solvent such as benzene, methylene chloride, n-hexane, cyclohexane, or diethyl ether (Barquet et al. 1981; Javada and Hajari 1986; Lu et al. 1975; McNeil et al. 1977). XAD-2 resin has also been used to sequester heptachlor from water samples, followed by elution with solvents such as diethyl ether (Green et al. 1986; Strachan 1985). Pesticide-solvent mixtures are concentrated and dried using anhydrous sodium sulfate and evaporation (Brodie et al. 1984). Methylene chloride is also used to partition the pesticide, followed by purification on an alumina or silica gel column (Ambrus et al. 1981b).

Fish are commonly used as monitors or indicators of pesticide contamination in aquatic environments. Fish or other aquatic organisms are ground and mixed to obtain a homogeneous sample (Bulkley et al. 1981; Teichman et al. 1978). Samples may be frozen prior to analysis. There are two common methods for extraction of pesticides from these organism homogenates. In the first, samples are mixed with petroleum ether and sodium sulfate (Erney 1983; Hashemy-Tonkabony and Langaroodi 1976; Henderson et al. 1969). Alternatively, a mixture of acetonitrile and water is added to the tissue samples, followed by phase separation centrifugation, and florisil or similar column chromatographic purification (Barquet et al. 1981; Henderson et al. 1969; Leung et al. 1981). Eluates are usually concentrated and then quantified.

Due to heptachlor's lipophilicity and bioconcentration potential, residues are often present in mammalian milk. To analyze contamination, sulfuric acid is slowly added to milk and mixed to separate curd particles from water. The organic layer is decolorized with additions of sulfuric acid, removed, and concentrated in an evaporator (Veierov and Aharonson 1980). Another method mixes milk with diethyl ether and hexane, followed by partitioning in acetonitrile and extraction into hexane (Gabica et al. 1974).

Plant materials are ground to homogeneity and may be frozen until analysis. Samples are mixed with acetone or methylene chloride and filtered. The sample may be dried with sodium sulfate or by steam distillation. Steam distillation has been shown to transform some of the heptachlor to hydroxychlordene which must be accounted for in the analysis (Nash 1984). Methylene chloride or 2,2,4-trimethylpentane are commonly used to partition the samples, followed by purification on alumina or silica gel columns (Ambrus et al. 1981b; Nash 1984).

Heptachlor or heptachlor epoxide residues in prepared samples are quantified, in most cases, by gas chromatography (GC) with an electron capture detector (Alford-Stevens et al. 1986; Henderson et al. 1969; Leung et al. 1981; McNeil et al. 1977). Nitrogen is the commonly used carrier gas, although argon-methane mixtures are also used (McNeil et al. 1977). Few reports of detection limits are found in the literature. Hashemy-Tonkabony and Langaroodi (1976), Veith et al. (1981), and Teichman et al. (1978) reported sensitivities of 0.001 $\mu g\ g^{-1}$, 0.50 ng g^{-1}, and 0.05 ng g^{-1} respectively, with gas-liquid chromatography.

Thin-layer chromatography may also be used as a simple and rapid method for the detection of heptachlor. Concentrated samples are spotted on alumina-$AgNO_3$ or aluminum oxide G prepared plates with n-heptane as a solvent (Ambrus et al. 1981a; Javadi and Hajari 1986). Sodium fluorescein or silver nitrate with UV irradiance are fairly specific visualization agents (Ambrus et al. 1981a; Feroz and Khan 1979; Javadi and Hajari 1986).

IV. Transport, Transformation, and Environmental Concentrations

The transfer of pesticides from their sites of application toward nontarget systems is a key detrimental result of the compound's use. A variety of physical, chemical, and biological factors affect these transfers (Fig. 2). The three conduits for transfer are the atmosphere, lithosphere (soil), and hydrosphere (aquatics); the biota interact with all three. Within each compartment, physicochemical and biochemical processes produce many degradation products and metabolites of heptachlor (Fig. 3).

Surface or aerial spray application of heptachlor does not purposely release the compound for dispersion via the atmosphere. Rather, any atmospheric transfer of heptachlor results from noncontainment of the spray medium, or from volatilization of the pesticide from the site of its terrestrial application. Of these two, volatilization can play a mayor role: nearly 50% of surface-applied heptachlor volatilized from a treated soil within 12 hr, and 90% within 2-3 d (Taylor et al. 1976, 1977). Soil factors governing heptachlor volatilization will be discussed later in this section.

In the air, a variety of phototransformations of methano-bridged cyclodiene insecticides occurs (Ivie et al. 1972). Though stable to visible light (Worthing 1979), sunlight or UV light produce complex heptachlor photoproducts, including dechlorination products and caged structures (Ivie et al. 1972; Knox et al. 1973; McGuire et al. 1972; Parlar et al. 1978; Podowski et al. 1979). Most photoisomers are probably of limited toxicological importance to nontarget organisms due to the low amounts produced (Ivie et al. 1972), even though the

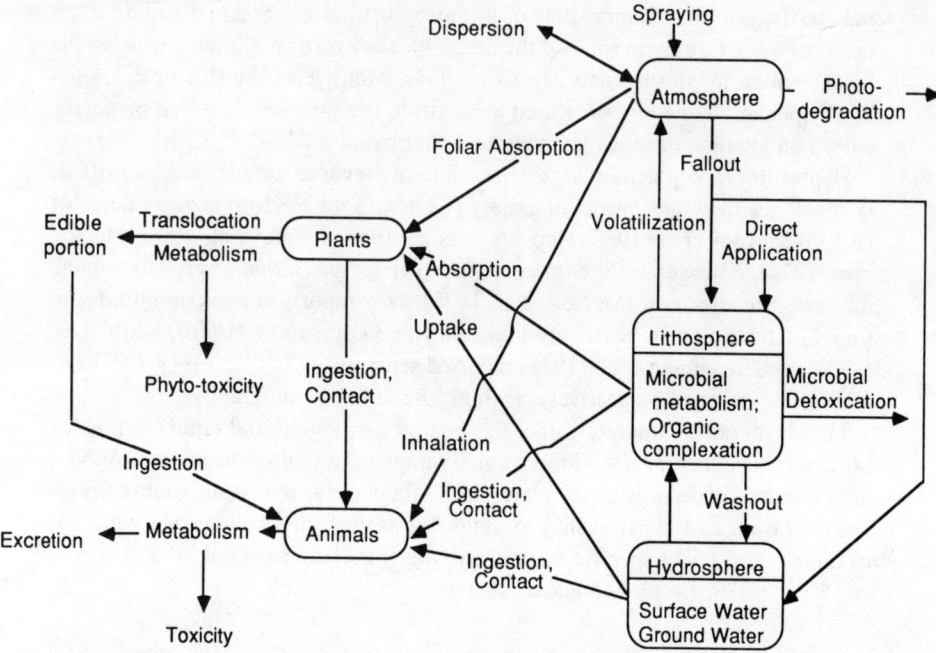

Fig. 2. Potential conduits for transport and transformation of heptachlor in the environment.

photoproducts may be more toxic than the parent compounds (Podowski et al. 1979; Rosen et al. 1969).

The mean "background" concentration of heptachlor in US air has been estimated at 0.5 ng heptachlor m^{-3} (Peirano 1980), while the air over the southwestern Atlantic ocean was recently found to contain 0.016–0.131 ng m^{-3} (Weber and Montone 1988). Heptachlor was detected in the air of two of nine US cities, at a maximum concentration of 19.2 ng m^{-3} (Stanley et al. 1971). In the Mississippi delta agricultural region, Arthur et al. (1976) reported maximum pesticide levels of 0.8 ng heptachlor m^{-3} and 9.3 ng heptachlor epoxide m^{-3} in 156 weekly air samples during 1972–1974. Atmospheric levels of heptachlor become quite elevated over background concentrations following agricultural application: immediately after treatment of agricultural fields with 2.24 kg technical grade heptachlor ha^{-1} (224 mg m^{-2}), the concentration of heptachlor in the air was 244 ng m^{-3}, and after 3 weeks, 15.4 ng m^{-3} (Peirano 1980).

Heptachlor in the soil is subject to multiple transformation and degradation reactions, intimately coupled with biotic and abiotic influences. On exposed soil and plant surfaces, heptachlor is potentially transformed by the photoreactions mentioned earlier. In the soil, heptachlor transformation occurs by at least three pathways: epoxidation, hydrolysis, and dechlorination (see Fig. 3) (Miles et al.

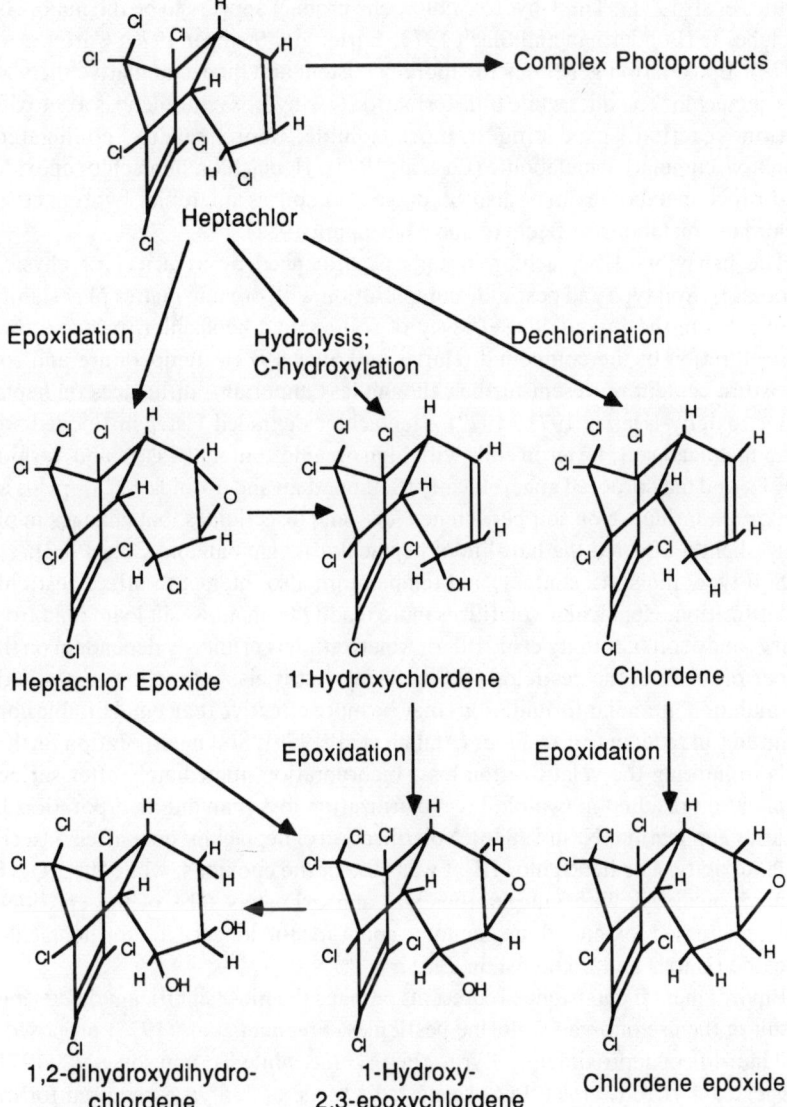

Fig. 3. Potential products and metabolites of heptachlor in the environment. (Modified from Miles et al. 1969)

1969). Heptachlor epoxidation in soils (Gannon and Bigger 1958) is predominantly microbially-mediated (Miles et al. 1969). Hydrolysis or C-hydroxylation to 1-hydroxychlordene, or bacterial dechlorination to chlordene are two other transformation pathways (Miles et al. 1969; Tashiro and Matsumura 1978; WHO 1984b). Heptachlor epoxide can be further metabolized to 1-hydroxychlordene

(Miles et al. 1971). The 1-hydroxychlordene product appears to be the major soil metabolite (Bonderman and Slach 1972; Carter and Stringer 1970; Carter et al. 1971). Epoxidation generates the more persistent and bioaccumulative metabolite heptachlor epoxide, while transformation to 1-hydroxychlordene is a detoxification reaction, producing a more soluble, more readily conjugated, nonbioaccumulative metabolite (Lu et al. 1975). Heptachlor, heptachlor epoxide, and other metabolites may also be present in soil as microbial byproducts of chlordane metabolism (Beeman and Matsumura 1981).

The behavior of heptachlor in soils is influenced by a variety of physical processes. Soil type and pesticide complexation with organic matter play significant roles in the insecticidal efficacy of soil-applied heptachlor, more so than concentration of the compound (Harris and Sans 1972). Temperature and soil moisture content represent further, though less important, influences on heptachlor toxicity (Harris 1971, 1972). Heptachlor degraded faster in flooded soil than in upland soil, faster in soils with high organic content (Castro and Yoshida 1971), and may proceed anaerobically (Sethunathan and Yoshida 1973). pH is an important influence on soil persistence for some insecticides, but changes in pH only slightly affected the half-life of heptachlor (Chapman and Cole 1982).

Soil type, moisture content, and temperature also interact to affect pesticide volatilization. Heptachlor volatilizes more readily from moist silt loam than from a dry sandy soil (Glotfelty et al. 1984). Volatization is primarily dependent on the vapor pressure of the pesticide (Nash 1983a), but is also a function of pesticide formulation: granular formulations may be more effective than emulsifiable concentrates in reducing air residues (Atallah et al. 1979). Soil incorporation further aids in limiting the volatilization loss. Incorporation immediately after surface application resulted in twofold less volatilization loss than did incorporation 12 hr after application (Nash 1983b). Volatilization of heptachlor does affect insecticidal toxicity. The heptachlor LT_{50}[1] was 20% of the epoxide's, while the LD_{50} in contact bioassays with vinegar flies was approximately 60% of the epoxide's; thus, decreased toxicity alone cannot account for the longer LT_{50} for heptachlor epoxide (Harris and Lichtenstein 1961).

Environmental persistence represents perhaps the most significant deleterious result of the use of organochlorine pesticides. Freeman et al. (1975) proposed a soil half-life of approximately 1 yr for heptachlor, while Hermanson et al. (1971) suggested 4 yr. Edwards (1966) suggested a range of 3-5 yr as required for the disappearance of 95% of an initial amount of applied heptachlor, while Turner et al. (1972) suggested 2-3 yr for 95% depletion of heptachlor, with some unknown longer time for heptachlor epoxide. These differences are most probably due to differing experimental conditions. Stewart and Fox (1971) found no detectable heptachlor residues in Nova Scotian soils 9-13 yr after application of 3.4, 4.5, or 6.7 kg heptachlor ha^{-1}; any residual compound was exclusively in the epoxide

---

[1]LT_{50} (median lethal time): the time required to cause lethality in 50% of a population exposed to a toxicant.

form. Sixteen years after extremely heavy rates of application (up to 224 kg ha^{-1}), residues, primarily heptachlor epoxide, were present in a sandy loam at a mean total concentration of 10% of the original application amount (Nash and Harris 1973). Soybeans grown on this soil 15 yr after application contained residues of heptachlor epoxide. Nine years following a single application of 5.6 kg ha^{-1}, no reinfestation by European wireworm of a continuously cultivated soil occurred, indicating persistent toxicity at residue concentrations of 9 ng g^{-1} and 169 ng g^{-1} for heptachlor and heptachlor epoxide, respectively (Wilkinson et al. 1964). Seven years after application of several insecticides to soil, the residual toxicity to termites was: dieldrin > aldrin > heptachlor > chlordane > DDT > lindane > sodium arsenite; while ranking on percentage of termiticide remaining was: DDT > dieldrin > chlordane > aldrin and converted aldrin > heptachlor and heptachlor epoxide > lindane > sodium arsenite (Bess and Hylin 1970).

Mobility of the pesticide is also an important concern. Movement of heptachlor through soils was directly correlated with topography (Peach et al. 1973). Three years after application, heptachlor was still present in a subtropical soil, but was almost exclusively confined to the upper 15 cm; however, there was some movement of heptachlor epoxide beyond this level (Talekar et al. 1983). Surface application resulted in only 3% of the applied heptachlor remaining after 3-4 mon, while 15% remained after 5 mon when heptachlor was incorporated into the soil (Saha and Stewart 1967). Disturbance of treated soil (cultivation) resulted in a 76-82% decrease in heptachlor residues compared to an alfalfa-covered non-cultivated soil also treated with heptachlor (Lichtenstein et al. 1971). Soil cultivation increases volatilization, while cover crops decrease losses. In addition, 26% of the recovered residues were in the upper 5 cm soil layer, 52% in the 5-10 cm layer, and 5-6.5% in the 15-23 cm layer, mostly in the form of heptachlor epoxide. Qualitatively similar results were reported earlier by Lichtenstein et al. (1962).

When registered for agricultural application, heptachlor usage was relatively limited. In 1971, heptachlor was applied to 0.5% (8 of 1,473) of the National Soils Monitoring Program cropland sites at an average total application of 1.42 kg ha^{-1} (Carey et al. 1978b). In 1972, heptachlor was applied to 0.4% (5 of 1,402) cropland sites at an average rate of 1.41 kg ha^{-1} (Carey and Gowen 1979). Furthermore, usage was restricted to certain crops, with 0.01-0.6 µg heptachlor epoxide g^{-1} soil in 24% of vegetable or cotton field soil samples, 0.002-0.09 µg g^{-1} in 15% of the small grain and root crop soils, and no residues reported for fruit tree soils (Stevens et al. 1970). Truhlar and Reed (1976) found 0.6 µg of heptachlor epoxide kg^{-1} in the top 7.5 cm of 'general farm' (sic) soil, with no residues in forest, residential, or orchard soils. However, Frank et al. (1976) reported the presence of heptachlor and heptachlor epoxide residues in orchard soils, even though neither had been recommended for use in orchards, suggesting some degree of extralabel use.

Soil residues are reported from a variety of sources. In several urban monitoring studies, the mean concentration of heptachlor and heptachlor epoxide was nearly exclusively reported as <0.01 µg g^{-1} (Carey et al. 1976, 1979b; Lang et al. 1979; Wiersma et al. 1972a). In numerous surveys of agricultural lands, heptachlor and heptachlor epoxide residues were detected in 1–24% of the soil samples, and most mean values were ≤ 0.01 µg g^{-1} (Carey et al. 1972, 1978a, 1979a; Gowen et al. 1976; Harris and Sans 1971; Saha and Sumner 1971; Trautman et al. 1968; Wiersma et al. 1972b). With application of 0.47 µg of heptachlor g^{-1} soil, average residual concentration was approximately 0.01 µg g^{-1} (Mullins et al. 1971). In 20 of 1346 (1.5%) sites surveyed, with an average application of 0.885 kg heptachlor ha^{-1}, residual heptachlor averaged 0.013 kg ha^{-1} (Crockett et al. 1974). Application of heptachlor in a Great Plains irrigation district resulted in total heptachlor and heptachlor epoxide residues of 0.16–1.30 µg g^{-1} in soil at harvest time, without successive yearly accumulations; about 95% of the combined residues disappeared within a year (Knutson et al. 1971). Though banned from use in 1974, heptachlor was detected in 2% of Taiwanese tea-garden soils in a 1977 survey, and no heptachlor epoxide, while in 1984, no heptachlor was detected but heptachlor epoxide was found in 1% of the samples (Wang et al. 1988).

Transfer of heptachlor and heptachlor epoxide from the atmosphere or lithosphere to the aquatic environment is the final and often significant link in the transport and transformation sequence. From a survey of 11 agricultural watersheds in Canada, pesticides were found to enter the stream water from storm runoff (60%), base flow from internal soil drainage (18%), or accidental "spills" (22%) (Frank et al. 1982). Of 81 pesticides used in these agricultural systems, heptachlor expoxide and four others were found in stream water at levels that exceeded the International Joint Commission criterion of 1 ng L^{-1}. Terrestrial insecticides such as heptachlor, reach aquatic systems by aerial spray drift, run-off from treated areas, industrial discharges, precipitation, and from accidental spillage. Run-off from agricultural usage probably presents the largest chronic impact to waterbodies, while accidental discharges are more serious, but localized.

Suspended solids collected over a 3-yr period from the mouths of Canadian surface waters contained 0.1–3.6 µg heptachlor epoxide kg^{-1}, but no heptachlor (Frank et al. 1981). Frank (1981) correlated pesticide usage in the Grand and Saugeen river basins in Canada to the concentrations in the water. While no specific data on heptachlor usage was available, the author noted that the levels of heptachlor epoxide present in the river systems could not be accounted for solely by the reported agricultural usage of chlordane (contaminated with heptachlor). The differences could be from the use of chlordane as a lawn insecticide, with the excess loading due to run-off from these residential surface treatments. In 1975–1976 and 1976–1977, the yearly mean concentrations of heptachlor epoxide in the Grand River were, respectively, 0.07 and 0.03 ng L^{-1}, and "not detected" (nd) and 0.05 ng L^{-1} from the Saugeen River. Heptachlor concentrations measured in US surface waters ranged from 0.1 to 6.3 ng L^{-1} (McEwen and Stephenson 1979). In other work, Peirano (1980) found levels of heptachlor in major US river basins ranging from 1 to 35 ng L^{-1}. Heptachlor was found in water from ten sites in northern Mississippi with concentrations ranging from 0.03 to 0.18 ng L^{-1} (Rihan et al. 1978). Bradshaw et al. (1972) identified two major surges of heptachlor and heptachlor epoxide into a Utah lake in late spring and late fall (coinciding with application periods), which produced concentrations as high as 2.9 µg L^{-1}. In various samples taken from surface waters in Germany, heptachlor concentrations ranged from 10 to 205 ng L^{-1}, while heptachlor epoxide levels ranged from 25 to 40 ng L^{-1} (Herzel 1972). Still, few of the 28 sampling sites had any detectable heptachlor residues. Samples from streams entering the Blanca Bay, Argentina, had concentrations of heptachlor ranging from 5–32 ng L^{-1} in filtered water and 9–390 ng g^{-1} on suspended matter (Zubillaga et al. 1987). Heptachlor residues as high as 460 ng L^{-1} have been found in waters in Nova Scotia (Burns et al. 1975).

In potable water supplies, heptachlor and heptachlor epoxide were found in approximately half of the samples tested in a South Carolina study (Sandhu et al. 1978). The ranges of heptachlor and heptachlor epoxide residues were nd-44 ng L^{-1} and nd-87 ng L^{-1}, respectively. Mean concentrations of heptachlor in rural drinking water ranged from 9–15 ng L^{-1}. Heptachlor and heptachlor epoxide were detected in samples of drinking water from Ottawa with mean concentrations of 0.6 and 3.0 ng L^{-1}, respectively (Williams et al. 1978). The WHO (1984a) recommends an upper limit of 100 ng L^{-1} for heptachlor and heptachlor epoxide in potable water.

Due to the then widespread agricultural use of heptachlor, Lu et al. (1975) followed its distribution and fate in a laboratory model ecosystem. Approximately 40% of the heptachlor was metabolized in fish to 1-hydroxychlordene. The other major metabolite was heptachlor epoxide. The epoxide is relatively inert and therefore, very stable, due to the absence of the activating allylic moiety of heptachlor. In a pair of concurrently run studies. Eichelberger and Lichtenberg (1971) found 25% of initially added heptachlor remained in river water after 2 wk and 0% after 4 wk, while 100% of the initially added epoxide was present after both 2 and 4 wk.

Seasonal variations may play a role in concentration profiles of heptachlor. In a Massachusetts study, Smith and Cole (1970) found that the residues of heptachlor and heptachlor epoxide in juvenile fish correlated better to the spring period of high run-off than to the late spring–summer pesticide application times. The peak in heptachlor epoxide concentration was also observed later than the peak for the parent compound heptachlor.

In rainwater analyses, relatively little parent compound or epoxide was present, while photometabolites such as 1-hydroxychlordene and 1-hydroxy-2,3- epoxychlordene predominated (Lu et al. 1975). Heptachlor epoxide residues were measured in the rainfall around Lake Superior in 1983 and 1984, where concentrations ranged from 0.03 to 1.3 ng L^{-1} at two sampling points (Strachan 1985, 1988). Measurement of the hydroxychlordene residues might have provided a more accurate picture of the concentrations of heptachlor breakdown products.

V. Ecological Toxicology

Terrestrial insects are the target organisms for heptachlor usage; however dispersion and persistence of the pesticide in the environment provides an opportunity for heptachlor exposure in terrestrial and aquatic nontarget species, and bioaccumulation through food-web transfers.

A. Terrestrial Biota

There are multiple pathways and mechanisms within the microbial community for degradation, transformation, and metabolism of heptachlor (Fig. 3 and section IV). Some of these further metabolites of heptachlor and heptachlor epoxide are not later degraded by microbial action, and may then accumulate to high levels in soil (Miles et al. 1969), but knowledge of cytological and biochemical effects of organochlorine insecticides on microorganisms is too limited to characterize conclusively these effects (Lal and Saxena 1982).

While microbial laboratory studies provide useful metabolic information, care must be exercised when extrapolating to field situations, since bacterial growth assays are dependent on experimental conditions (Collins and Langlois 1968). Heptachlor inhibited the growth of a wide range of gram-positive bacteria but not gram-negative bacteria; the selectivity was attributed to blockage of electron transfer and/or interference with cell wall or membrane processes (Audus 1960). Heptachlor at amounts ordinarily used in field practices inhibited the growth of 50% of the soil fungi, 81% of the actinomycetes, and 89% of bacteria isolated from field soil samples (Shamiyeh and Johnson 1973). Higher concentrations of heptachlor caused population increases in the residual resistant bacteria, perhaps from reduced competition. Heptachlor increased respiration and decreased reproduction in *Staphylococcus aureus* colonies (O'Neill and Langlois 1976). Decreases in *Nitrobacter* reproduction also resulted from heptachlor exposure; however, there was no change in NO_2^- oxidation, the key ecological role of the genus (Rennie 1977). Application of heptachlor to soils at agricultural rates had no harmful effect on nodulation and yield of groundnut (*Arachis hypogaea*) inoculated with *Rhizobium* (Kulkarni et al. 1974). Heptachlor inhibited the growth of yeast, seemingly from impaired oxidative metabolism (or component synthesis) (Nelson and Williams 1971).

Plants are exposed to heptachlor through direct foliar contact, root uptake, or foliar absorption of vapors (Beall and Nash 1971). Root absorption was the major contamination route in soybeans for heptachlor, endrin, and dieldrin (Beall and Nash 1971; Nash and Beall 1970). In comparison among several organochlorines, the relative magnitudes of foliar absorption for DDT, dieldrin, endrin, and heptachlor were equivalent, approximately 6.5 µg of pesticide g^{-1} plant (Beall and Nash 1971).

Heptachlor treatment does not seem to affect plant growth adversely, though the data of Gawaad et al. (1972) showed varied effects on germination, root and stem growth, and plant matter dry weight for cotton, corn, clover, and bean when soils were sprayed with a solution of 0–50 µg of heptachlor ml^{-1} acetone. Seed treatment with heptachlor did not reduce seedling emergence in three wheat cultivars (Khaleeq and Klatt 1986), nor did heptachlor cause any discernible phytotoxicity or reduction of stem elongation in alfalfa (Fick 1977).

Heptachlor epoxidation occurred on certain plants, and may be common to all plants (Gannon and Decker 1958). A 1.12 kg ha^{-1} treatment of heptachlor on alfalfa yielded heptachlor residues which persisted above 0.1 µg g^{-1} for 13 d, while the epoxide did not reach this level until 25 d after treatment. The heptachlor epoxide:heptachlor ratio is generally much greater in soybean tissues than in the soil (except in the youngest plants, where equal amounts of the two materials were found), suggesting that considerable epoxidation occurred in soybean tissues (Turner et al. 1972). Lower leaves contained the largest concentration of heptachlor epoxide; this may reflect uptake of heptachlor that had volatilized from the soil. In a study utilizing soil treated with 0.5 µg of [^{14}C]-heptachlor g^{-1}, heptachlor epoxide was present in soybeans at < 0.1 µg g^{-1} in hay and < 0.04 µg g^{-1} in seed; no residues were detected in cotton (Nash et al. 1970).

The transfer of pesticide residues from soil to plant is an important trophic concern. In soils treated with 5.6 or 11.2 kg heptachlor ha^{-1} (a typical agricultural treatment is 2.2 kg ha^{-1}), rutabagas contained approximately 0.01 µg of heptachlor g^{-1} plant, and 0.03 µg of heptachlor epoxide g^{-1}; carrots contained 0.26 µg of heptachlor g^{-1} and 0.04 µg of heptachlor epoxide g^{-1}, with an accompanying off-flavoring noted (Fox et al. 1964). Comparison of heptachlor uptake from sandy soils or from muck soils showed that sandy soils produce quantifiably more insecticide in carrots, though the levels of nd to <0.05 µg g^{-1} did not seem to pose a hazard to humans (Oloffs et al. 1971). Heptachlor was detected in alfalfa planted on cornfield soil previously treated with heptachlor (Moubry et al. 1967). Heptachlor was not detected in foliage, though heptachlor epoxide was occasionally found at <0.01µg g^{-1} in corn harvested from soil containing residues of 0.16–1.30 µg g^{-1} (Knutson et al. 1971).

Three years after cessation of heptachlor application at a total of 33.5 kg ha^{-1} added over a 3-yr period, residues were analyzed in soils and a resident potato crop. Soils contained 2.1 µg of γ-chlordane g^{-1}, 2.0 µg of heptachlor g^{-1}, and 0.8 µg of heptachlor epoxide g^{-1}, while potatoes contained 0.016 µg of γ-chlordane g^{-1}, 0.017 µg of heptachlor g^{-1}, and 0.06 µg of heptachlor epoxide g^{-1} (Stewart et al. 1965). In addition, half of the total residues were present in or on the surface layer of the potato tuber. At an application rate of 2 kg ha^{-1}, heptachlor and heptachlor epoxide residues in a potato crop exceeded a prescribed maximum residue limit of 0.05 µg g^{-1} (Chawla et al. 1981). In these plants, insecticide absorption increased 300–500% when potatoes were grown in sandy soils, resulting from the intrinsically lower ability of sandy soils to retain the pesticide.

In tobacco plants grown on heptachlor-treated soils, both heptachlor and its epoxide were detected in all strata of tobacco leaves ('sand', middle, and top)

(Townsend and Specht 1975). Similarily, corn leaves 60 cm aboveground contained approximately fivefold greater heptachlor residues than leaves 180 cm aboveground (Caro 1971). With wheat seed treated with 521 or 1,042 μg of heptachlor g^{-1}, there were no detectable heptachlor or heptachlor epoxide residues in the harvested grain (Burrage and Saha 1967).

Most data regarding heptachlor and heptachlor epoxide in animals consist of residue reports, frequently analyzed in conjunction with other organochlorines. In some cases, biomagnification can be demonstrated (Hernandez et al. 1987). Table 3 compiles much of the observational data for terrestrial vertebrates. While residue reports provide useful information on pollutant dispersion and environmental accumulation, it is often very difficult to assess accurately exposure to the compound, especially for long-lived or mobile species. A second difficulty in interpreting residue levels lies in attributing a specific effect or response to a specific residue; this problem is compounded when a host of residues are quantified, and significance is thereby determined statistically, not mechanistically.

Efforts in analysis of the effects of heptachlor on terrestrial invertebrates have centered on the potential transfer of soil residues through higher-order animals via invertebrate prey (cf. section V.C). From one field study, heptachlor epoxide occurred in nine of 67 soil samples, and heptachlor, heptachlor epoxide and γ-chlordane were detected in earthworms collected from these soils (Gish 1970), but only four samples exceeded 0.1 μg g^{-1}. Residues in this range of concentrations are probably not toxic to earthworm-feeding birds (cf. section V.C). Heptachlor and heptachlor epoxide are more toxic to earthworms than aldrin and dieldrin, but all are equitoxic to soil arthropods (Edwards and Thompson 1973).

Studies concerning effects of heptachlor on amphibians are sparse. One physiological study revealed that heptachlor had no significant effect upon sodium active transport and ionic permeability of isolated frog (*Rana* sp.) skin, in contrast to other organochlorines (Webb et al. 1979). Environmentally, in a lake treated with technical chlordane, California newts (*Taricha torosa*) epoxidized all bioaccumulated heptachlor within 14 d, and eliminated all heptachlor epoxide within 9 mon (Albright et al. 1980). DeWitt et al. (1960) presented a study where agricultural application of 2.2 kg of heptachlor ha^{-1} produced elevated tissue residues and associated mortality of several amphibian species.

Reptiles have received somewhat more attention regarding the environmental effects of pesticides. Heptachlor epoxide was present in samples from five of eight egg clutches in nests of the endangered American crocodile (*Crocodylus acutus*) at Everglades National Park (Hall et al. 1979). It seemed unlikely that these residues adversely affected reproduction. Using the American alligator as a biomonitor of various environmental pollutants, Delany et al. (1988) provided a preliminary recommendation that pollutant load (including heptachlor residues) in alligator meat did not pose a threat to the meat consumer, at residue levels of 2.39 ± 1.52 μg heptachlor g^{-1} lipid weight (lw). Organochlorine residues were found in loggerhead (*Caretta caretta*) and green turtle (*Chelonia mydas*) eggs, but were well below any apparent harmful level (Clark and Krynitsky 1980). Mortality of two species of snakes (*Natrix* sp., *Heterodon* sp.) was associated with agricultural application of 2.2 kg heptachlor ha^{-1} (DeWitt et al. 1960). For a review of the effects of pesticides (including heptachlor) on reptiles, see Hall (1980).

The most common end-point in residue analysis of bird tissues and eggs is an attempt to relate organochlorine contamination with impaired reproductive success, via assessment of eggshell thinning (as in the case of DDT). Thus, heptachlor and heptachlor epoxide data have often been but one further component of studies analyzing a host of organochlorine contaminants; as such, direct specific evidence for impairment of bird reproductive function by heptachlor and heptachlor epoxide is scarce. Egg analyses indicate the presence of heptachlor and heptachlor epoxide (Table 3), but residues are generally low enough as to not cause any obvious direct impairment of reproductive function or success, as found by Szaro et al. (1979) with several seabird species. Again, while a general association between high organochlorine levels and low breeding success may be established (cf. De Smet 1987; Elliot et al. 1988; Pearce et al. 1979; Teeple 1977), direct evidence for a primary role of heptachlor and heptachlor epoxide in this impairment is lacking. Heptachlor in fact, may not play a role in reproductive impairment. The data of Elliot et al. (1988) showed a decrease in DDT-type compounds, PCBs, and several other organochlorines and a concommitant rise in breeding success and population numbers of northern gannets (*Sula bassanus*) from 1968 to 1984, but heptachlor epoxide residues remained constant (and α-HCH appeared to increase).

To identify potential organochlorine-induced lethality in wild birds, Wiemeyer and Cromartie (1981) suggested using carcass organochlorine content to predict brain concentration. When coupled with the lethal brain concentration estimates of Stickel et al. (1979) (3.4–8.3 µg of heptachlor epoxide g^{-1} brain tissue), organochlorine poisoning may be more readily identified.

LC$_{50}$ values (concentration of heptachlor in food) for captive bobwhite quail, Japanese quail, ring-necked pheasants and mallards in 5 d feeding tests were in the range of 92–480 µg of heptachlor g^{-1} food, wet weight (ww) (Hill et al. 1975). Two quail were killed after field application of <2.2 kg technical grade heptachlor ha^{-1} (Clawson and Baker 1959). Heptachlor epoxide toxicity was strongly implicated in the death of a merlin (*Falco columbarius*), which died after mobilization of the epoxide from fat reserves during migration (Henny et al. 1976). Heptachlor epoxide and oxychlordane may have been involved in the mortality of long-billed curlews (*Numenius americanus*) from Oregon (Blus et al. 1985a). Chlordane was potentially responsible for deaths of two adult female red-shouldered hawks (*Buteo lineatus*), but an interaction between heptachlor epoxide, dieldrin, and chlordane may also have been important (Blus et al. 1983). Elevated total organochlorine residues (including heptachlor epoxide) may have caused the death of a great horned owl (*Bubo virgianus*) from New York

Table 3. Heptachlor and heptachlor epoxide residues in terrestrial vertebrates

Organism Class Order Family Genus–species (common name)	n/N	Source (Tissue)	$\bar{x} \pm sd$	Range	Reference
Amphibia					
Anura					
Ranidae					
Rana spp. (frog)		WB		nd-20.9	DeWitt et al. 1960, Dowd et al. 1985
Bufonidae					
Bufo sp. (toad)	1/	WB	3.1		DeWitt et al. 1960
Reptilia					
Squamata					
Colubridae					
Natrix spp. (water snakes)		WB		nd-11.3	DeWitt et al. 1960, Dowd et al. 1985
Heterodon sp. (hognosed snake)	1/	WB	4.2		DeWitt et al. 1960
Crocodilia					
Crocodylidae					
Crocodylus acutus (American crocodile)	5/8	Eggs	0.01 ± 0.00	tr-0.04	Hall et al. 1979
Alligatoridae					
Alligator mississippiensis (American alligator)	8/8	Tail muscle	2.39 ± 1.52(lw)	0.8-5.0(lw)	Delaney et al. 1988

Aves

Podicipediformes					
Podicipedidae					
Podiceps grisegena (red-necked grebe)	10/	Eggs	1.79 ± 1.12(lw)	0.30–3.57(lw)	De Smet 1987
Pelecaniformes					
Pelecanidae					
Pelecanus erythrorhynchos (white pelican)	9/	Eggs	0.05		Greichus et al. 1973
	3/	Fat	1.33		Greichus et al. 1973
		Several		tr–13.20	Benson et al. 1976, Greichus et al. 1973
Pelecanus occidentalis (brown pelican)		Several		nd–0.26	Blus et al. 1977, 1979
	56/580	Eggs	0.16 ± 0.05	nd–0.53	Blus et al. 1977, 1979
Sulidae					
Sula bassanus (northern gannet)	24/	Eggs (1968–74)	0.046 ± 0.008	0.037–0.053	Elliot et al. 1988
	30/	Eggs (1969–84)	0.032 ± 0.004	0.029–0.040	Elliot et al. 1988
Phalacrocoracidae					
Phalacrocorax auritus (double-crested cormorant)		Eggs		tr–0.06	Weseloh et al. 1983, Zitko and Choi 1972
	17/	Eggshell	0.05		Greichus et al. 1973
	8/	Fat	0.25		Greichus et al. 1973
	10/	Several	<0.05		Greichus et al. 1973
Phalacrocorax olivaceus (olivaceous cormorant)	4/8	Carcass		tr–0.02	King and Krynitsky 1986
	11/17	Eggs		tr–0.32	King and Krynitsky 1986, Morrison et al. 1978

Table 3. Continued

Organism Class Order Family Genus – species (common name)	n/N	Source (Tissue)	$\bar{x} \pm sd$	Range	Reference
Aves *(continued)*					
Ciconiiformes					
Ardeidae					
Ardea herodias	14/	Brain	0.51 ± 0.52	0.13–1.9	Ohlendorf et al. 1981
(great blue heron)	11/	Carcass	0.34 ± 0.21	0.11–0.80	Ohlendorf et al. 1981
Bubulcus ibis	2/	Brain		0.19–0.54	Ohlendorf et al. 1981
(cattle egret)	1/	Carcass	0.65		Ohlendorf et al. 1981
Florida caerulea	28/	Eggs		nd–0.15	Greenberg and Heye 1971
(little blue heron)		Several		nd–0.25	Greenberg and Heye 1971
Nyctanassa violacea (yellow-crowned night heron)	/5	WB	nd		Dowd et al. 1985
Nycticorax nycticorax (black-crowned night heron)	1/	Brain	0.36		Ohlendorf et al. 1981
	3/	Carcass	0.38 ± 0.16	0.22–0.53	Ohlendorf et al. 1981
Anseriformes					
Anatidae					
Anas acuta	2/3	Brain		nd–1.78	Flickinger and Krynitsky 1987
(pintail)	/14	Carcass		nd–5.0	Flickinger and Krynitsky 1987
Anas carolinensis	4/4	Brain		0.65–2.0	Flickinger and Krynitsky 1987
(green-winged teal)	6/10	Carcass		nd–9.3	Flickinger and Krynitsky 1987

Toxicology of Heptachlor

Species	Sample	Tissue	Value	Reference
Anas platyrhynchos (mallard)	1/14	Carcass	nd-0.08	Flickinger and Krynitsky 1987
	7/13	Eggs	nd-13.60	Blus et al. 1985b
		Wings	nd-1.7	Cain 1981, Fleming and Cain 1985, Heath 1969, Heath and Hill 1974, White 1979b
Anas rubripes (black duck)		Eggs	tr-0.16	Haseltine et al. 1980, Longcore and Mulhern 1973
		Wings	tr-0.06	Cain 1981, Heath 1969, Heath and Hill 1974, White 1979b
Aythya valisineria (canvasback)	2/97	Eggs	tr-0.85	Stendell et al. 1977
Branta canadensis (Canada goose)	6/113	Carcass	0.11 ± 0.00(se)	White et al. 1979
	/28	Brain	nd-49	Blus et al. 1984
	/216	Eggs	nd-24.29	Blus et al. 1984
Bucephala clangula (common goldeneye)	25/	Fat (1894)	0.03–0.41	Foley and Batcheller 1988
	24/	Fat (1985)	0.19–0.27	Foley and Batcheller 1988
Mergus merganser (common merganser)	2/2	Eggs	0.44	Haseltine et al. 1981
Mergus serrator (red-breasted merganser)	200/206	Eggs	nd-0.88	Haseltine et al. 1981
Falconiformes				
Cathartidae				
Aegypius monachus (black vulture)	1/	WB	0.008	Bijleveld et al. 1979
Accipitridae				
Accipiter cooperii (Cooper's hawk)	1/1	Muscle	0.002	Havera and Duzan 1986
Accipiter gentilis (goshawk)	1/7	Liver	1.5	Havera and Duzan 1986
Accipiter nisus (sparrow hawk)		Several	0.014–0.035	Sierra et al. 1987
		Fat	0.285	Sierra et al. 1987

Table 3. *Continued*

Organism Class Order Family Genus–species (common name)	n/N	Source (Tissue)	$\bar{x} \pm sd$	Range	Reference
Aves *(continued)* Falconiformes *(continued)* Accipitridae *(continued)*					
Aquila chrysaetos (golden eagle)	9/10 /32	Eggs Fat Several	0.007 ± 0.006	nd-0.02 <0.01-2.4 tr-12.2	Lockie and Ratcliffe 1964 Reidinger and Crabtree 1974 Reichel et al. 1969, Reidinger and Crabtree 1974
Buteo jamaicensis (red-tailed hawk)	5/	Eggs Several		0.09-0.80 tr-0.84	Seidensticker and Reynolds 1971 Havera and Duzan 1986, Seidensticker and Reynolds 1971
Buteo lagopus (rough-legged hawk)		Several		0.04-0.29	Havera and Duzan 1986
Buteo lineatus (red-shouldered hawk)	1/1 1/2	Fat Muscle	1.42 0.10		Havera and Duzan 1986 Havera and Duzan 1986
Haliaeetus leucocephalus (bald eagle)		Brain		tr-2.1	Barbehenn and Reichel 1981, Belisle et al. 1972, Cromartie et al. 1975, Kaiser et al. 1980, Mulhern et al. 1970, Prouty et al. 1977, Reichel et al. 1969

Toxicology of Heptachlor

Species		Tissue	Concentration	Reference
		Carcass	tr-20.0(lw)	Barbehenn and Reichel 1981, Belisle et al. 1972, Cromartie et al. 1975, Kaiser et al. 1980, Mulhern et al. 1970, Prouty et al. 1977, Reichel et al. 1969
Milvus milvus (red kite)	/2	Eggs	tr-0.08	Krantz et al. 1970, Wiemeyer et al. 1972
		Brain	0.056	Krantz et al. 1970, Wiemeyer et al. 1972
Parabuteo unicinctus (Harris's hawk)	4/5	Eggs	0.034–0.078	
			0.06 0.02–0.13	Mader 1977
Pandionidae				
Pandion haliaetus (osprey)	13/33	Brain	0.16 ± 0.18 0.02–0.41	Weimeyer et al. 1980
	16/56	Carcass	tr-0.38	Weimeyer et al. 1980, 1987
	4/11	Eggs	0.004 ± 0.002 nd-0.007	Littrell 1986
Falconidae				
Falco columbarius (merlin)		Eggs	1.7 (dw) tr-2.16 (dw)	Fyfe et al. 1976
Falco eleonorae (Eleonora's falcon)	/13	Eggs	0.0	Ristow et al. 1980
Falco mexicanus (prairie falcon)	2/4	Brain	5.05	Fyfe et al. 1969
		Eggs	1.3 (dw) tr-1.51 (dw)	Fyfe et al. 1976
Falco peregrinus (peregrine falcon)		Eggs	nd-0.24	Keck et al. 1982 (includes dieldrin residues)
Falco tinnunculus (kestrel)	/4	Fat	2.040 0.284–3.795	Sierra et al. 1987
	/4	Several	tr-2.042	Sierra et al. 1987

Table 3. Continued

Organism Class Order Family Genus – species (common name)	n/N	Source (Tissue)	$\bar{x} \pm sd$	Range	Reference
Aves *(continued)*					
Galliformes					
Phasianidae					
Calipepla californica (California quail)	1/1	Egg	2.95		Blus et al. 1985b
	1/1	Brain	15.0		Blus et al. 1985b
Colinus virginianus (bobwhite quail)	98/160	Wings		0.08–9.50	Montz et al. 1983
Phasianus colchicus (ring-necked pheasant)	64/	Brain		nd–21.0	Blus et al. 1985b, Linder and Dahlgren 1970
	72/	Eggs		nd–8.10	Blus et al. 1985b, Linder and Dahlgren 1970
	96/	Fat		nd–0.66	Blus et al. 1985b, Linder and Dahlgren 1970
Tetraonidae					
Pedioecetes phasianellus (sharp-tailed grouse)	46/	Fat	0.02	nd–0.16	Greichus et al. 1968
Gruiformes					
Rallidae					
Fulica americana (American coot)	/6	Several	<0.01		Greichus et al. 1978

Taxon	n	Tissue	Mean	Range	Reference
Charadriiformes					
Charadriidae					
Charadrius vociferus (killdeer)	12/19	Carcass	0.18 ± 0.11(se)	0.05–2.19	DeWeese et al. 1986
Scolopacidae					
Calidris alba (sanderling)	14/15	Carcass	0.19	0.1–0.3	White et al. 1980
Calidris pusilla (least sandpiper)	2/30	Carcass	0.32	0.2–0.5	White et al. 1980
Numenius americanus (long-billed curlews)	3/	Brain		1.0–4.8	Blus et al. 1985a
	7/	Eggs		nd–1.2	Blus et al. 1985a
Philohela minor (American woodcock)		Several		nd–0.62	Clark and McLane 1974, Edwards et al. 1983, McLane et al. 1971, 1978
Tringa flavipes (lesser yellowlegs)	600/	Wings	0.37 ± 0.32	tr–1.23	McLane et al. 1978
	2/17	Carcass	0.14	0.1–0.2	White et al. 1980
Laridae					
Larus argentatus (herring gull)		Eggs		nd–0.60	Bijleveld et al. 1979, Ellenton et al. 1985, Gilman et al. 1977, Norstrom et al. 1978
Larus atricilla (laughing gull)	69/	Liver	0.13 ± 0.05	0.04–0.24	Fox et al. 1988
	2/	Carcass		tr–0.2	King and Krynitsky 1986
	4/	Eggs		tr–0.5	Blus and Lamont 1979, King and Krynitsky 1986
Larus audouinii (Audouin's gull)	7/	WB		nd–0.031	Bijleveld et al. 1979
Larus delawarensis (ring-billed gull)	2/2	Brain		0.4–3.6	Blus et al. 1985b

Table 3. *Continued*

Organism Class Order Family Genus – species (common name)	n/N	Source (Tissue)	$\bar{x} \pm sd$	Range	Reference
Aves *(continued)*					
Charadriiformes *(continued)*					
Laridae *(continued)*					
Larus pipixcan (Franklin's gull)	6/	Several	<0.01		Greichus et al. 1978
Larus ridibundus (black-headed gull)	4/5	Fat Several		nd-37.03 nd-0.65	Vannucchi et al. 1978 Vannucchi et al. 1978
Rhyncopidae					
Rhynchops niger (black skimmer)	2/	Eggs		nd-0.1	King and Krynitsky 1986
Columbiformes					
Columbidae					
Columbia livia (rock dove)	1/3	Carcass	0.03 ± 0.03(se)	tr-0.08	DeWeese et al. 1986
Zenaidura macroura (mourning dove)		Several		tr-3.08	DeWeese et al. 1986, Edwards et al. 1983, Kreitzer 1974
Strigiformes					
Tytonidae					
Tyto alba (barn owl)	/23	Several		0.009-1.326	Sierra and Santiago 1987

Strigidae				
Asio flammeus (short-eared owl)	1/1	Heart	0.06		Havera and Duzan 1986
Asio otus (long-eared owl)		Several		0.01–0.97	Havera and Duzan 1986
Bubo virginianus (great-horned owl)	7/	Eggs		tr–0.23	Seidensticker and Reynolds 1971, Springer 1980
	/14	Fat	1.59	0.15–3.04	Havera and Duzan 1986
		Several		tr–4.0	Havera and Duzan 1986, Seidensticker and Reynolds 1971, Stone and Okoniewski 1983
Otus asio (screech owl)	1/4	Fat	0.11		Havera and Duzan 1986
		Several		0.02–0.83	Havera and Duzan 1986
Strix varia (barred owl)	2/6	Fat	5.48	0.80–10.16	Havera and Duzan 1986
		Several		0.01–1.40	Havera and Duzan 1986
Passeriformes					
Hirundinidae					
Petrochelidon pyrrhonota (cliff swallow)	3/14	Carcass	0.02 ± 0.01(se)	0.05–0.17	DeWeese et al. 1986
Hirundo rustica (barn swallow)	2/2	Carcass	0.06 ± 0.01(se)	0.06–0.07	DeWeese et al. 1986
Iridoprocne bicolor (tree swallow)		Carcass		0.02–0.29	DeWeese et al. 1986, Shaw 1984
		Eggs	0.06 ± 0.04		Shaw 1984
Tachycineta thalassina (violet-green swallow)	4/10	Carcass	0.12 ± 0.06(se)	0.05–0.46	DeWeese et al. 1986
Corvidae					
Corvus brachyrhyncos (common crow)		Several		tr–0.10	Greichus et al. 1978
Pica pica (black-billed magpie)	4/4	Brain	11.90 ± 5.15	4.6–16.0	Blus et al. 1985b
	9/9	Eggs	1.02	0.26–3.23	Blus et al. 1985b

Table 3. *Continued*

Organism Class Order Family Genus–species (common name)	n/N	Source (Tissue)	$\bar{x} \pm sd$	Range	Reference
Aves *(continued)*					
Strigiformes *(continued)*					
Turdidae					
Turdus migratorius (American robin)	17/35	Several		tr-0.968	DeWeese et al. 1986, Edwards et al. 1983
Sturnidae					
Sturnus vulgaris (European starling)		Carcass		nd-0.89	Greichus et al. 1978, Martin 1969, Martin and Nickerson 1972, Nickerson and Barbehenn 1975, White 1976, 1979a
	13/19	Eggs	0.21	nd-2.34	Blus et al. 1985b
Icteridae					
Agelaius phoeniceus (red-winged blackbird)	1/5	Carcass	0.07 ± 0.07(se)	tr-0.36	DeWeese et al. 1986
Sturnella neglecta (western meadowlark)	1/8	Carcass	0.01 ± 0.01(se)	tr-0.05	DeWeese et al. 1986
Euphagus cyanocephalus (Brewer's blackbird)	1/25	Carcass	0.01 ± 0.01(se)	tr-0.15	DeWeese et al. 1986

Mammalia					
Artiodactyla					
Antilocapridae					
Antilocapra americana (pronghorn antelope)	25/45	Renal fat		<0.03–0.12	Moore et al. 1968
Bovidae					
Oreamnos americanus (mountain goat)	13/	Renal fat	0.12 ± 0.09	0.02–0.28	Boddicker et al. 1971
Carnivora					
Canidae					
Canis lupus (timber wolf)	51/	Tongue and muscle	nd		Schneeweis et al. 1974
Vulpes vulpes (red fox)	9/10	Liver	27.0	9.4–90.6	Blackmore 1963
Mustelidae					
Lutra canadensis (river otter)	88/158	Liver		nd–0.003	Somers et al. 1987
Procyonidae					
Procyon lotor (Raccoon)	/5	Leg muscle	nd		Dowd et al. 1985
Chiroptera					
Vespertilionidae					
Myotis grisescens (gray bat)	50/	Brain		0.001–0.110	Kruthanut 1986
	50/	Carcass		nd–4.5	Clark et al. 1980, 1983
	8/	Carcass		36–375(lw)	Clark et al. 1980, 1983
	4/	Milk		nd–8	Clark et al. 1981
				nd–20	Clark et al. 1983

Table 3. Continued

Organism Class Order Family Genus–species (common name)	n/N	Source (Tissue)	$\bar{x} \pm sd$	Range	Reference
Mammalia *(continued)*					
Lagomorpha					
Leporidae					
Lepus europaeus (European hare)	199/	Liver fat	0.20(lw)	<0.01–1.38(lw)	Rimkus and Wolf 1987
Rodentia					
Cricetidae					
Reithrodontomys spp. (harvest mouse)				0.003–0.06	US Department of Agriculture 1969
Muridae					
Mus musculus (house mouse)				nd–0.02	US Department of Agriculture 1969
Sciuridae					
Citellus spp. (ground squirrel)			0.31		US Department of Agriculture 1969
Sciurus carolinensis (eastern gray squirrel)	17/22	Omental fat	0.02 ± 0.015(lw)	nd–0.058(lw)	Nalley et al. 1978

[a] All values are μg g^{-1} wet weight, mean ± standard deviation ($\bar{x} \pm sd$), unless otherwise noted (dw = dry weight; lw = lipid weight). Other abbreviations include: WB = whole body; nd = not detected; tr = trace; and (se) = standard error of the mean. The sample size (n/N) is the number of heptachlor or heptachlor expoxide-positive samples and the total number of samples. 'Several', as listed under Source (Tissue) is stated when the values are derived from many tissues, but not eggs nor fat.

(Stone and Okoniewski 1983). Some 36 bird species were found dead within 3 wk of exposure to agriculturally applied heptachlor (2.2 kg ha^{-1}) and all contained high levels of heptachlor epoxide (DeWitt et al. 1960). Lastly, agricultural heptachlor application at low rates may have caused less insect pest mortality (in this case, the fire ant *Solenopsis saevissima*), and resulted in contamination of insect prey at low levels, chronically exposing avian predators to heptachlor (Ferguson 1964).

With regard to potential impacts of heptachlor, heptachlor epoxide, and other organochlorines upon specific bird groups, raptors (predatory birds) receive considerable attention due to their position at the top of terrestrial food webs (cf. section V.C.). Deaths of several British peregrine falcons (*Falco peregrinus*), both wild and domestic, and domestic hunting lanners (*F. biarmicus*) probably resulted from the combined effects of dieldrin and heptachlor epoxide (Jeffries and Prestt 1966). Heptachlor and heptachlor epoxide caused direct mortality and decreased productivity in the American kestrel (*F. sparverius*) (Henny et al. 1983), with the kestrel being approximately seven times more sensitive to heptachlor and heptachlor epoxide than the Canada goose (*Branta canadensis*).

In waterfowl, lowered reproductive success and mortality of resident adult western Canada geese (*Branta canadensis moffitti*) were strongly associated with the use of heptachlor-treated wheat seed (Blus et al. 1984). Migratory waterfowl acquire the major portion of their organochlorine contamination along the migratory route or in their southern wintering grounds (Anderson et al. 1984; De Smet 1987; Foley and Batcheller 1988; Mora et al. 1987), but are believed to eliminate all but the PCBs and DDE annually (Anderson et al. 1984).

The woodcock (*Philohela minor*) had received some attention because of its gamebird status. Canadian woodcock were exposed to heptachlor and heptachlor epoxide on both wintering and breeding grounds, and young were already contaminated when hatched; thus, use of the pesticide coupled with concurrent DDT use posed potential problems for game management of that resource (Wright 1965).

In ecological studies, bobwhite quail populations (and other songbird populations) were reduced significantly following field application of heptachlor (Rosene 1965), while application of <2.2 kg technical grade heptachlor ha^{-1} caused a disappearance of all quail from an experimental site (Clawson and Baker 1959). A population decline of resident western Canada geese (*Branta canadensis moffitti*) was strongly associated with the use of heptachlor-treated wheat seed (Blus et al. 1984).

For mammals, as with birds, most studies report heptachlor and heptachlor epoxide residue levels, though the data are less numerous (Table 3). Eight species of mammals died within 3 wk of agricultural application of 2.2 kg heptachlor ha^{-1}, and mortality was correlated with elevated heptachlor residues (DeWitt et al. 1960). Nalley et al. (1978) found low and comparable heptachlor and heptachlor epoxide residues in park and residential gray squirrels (*Sciuris carolinensis*), with no apparent age, sex, or land-use influence on residue levels. Heptachlor residues (and metabolites of heptachlor) in bats were found to increase from 1976 to 1977, probably reflecting a switch to heptachlor as the pesticide-of-choice for cutworm control on Missouri farms (Clark et al. 1980, 1983).

Physiologically, heptachlor significantly elevated the ratio of free to protein-bound 11-hydroxycorticosteroids in "sexually mature, wild house mice" (sic) fed heptachlor in the diet at "agricultural doses" (sic) (Kamenov and Zolotarev 1979). The authors concluded that heptachlor was probably a strong stress factor, in addition to being a basic metabolic poison.

In a lone ecological study, there was no apparent influence of pesticide application (at normal agricultural rates) upon the dynamics of several rodent species populations living adjacent to treated fields, though pesticide residues were detected in rodent tissues (Robel et al. 1972).

B. Aquatic Animals

Table 4 summarizes data on the acute toxicity of heptachlor to aquatic organisms.

Exposure of the marine dinoflagellate *Exuviella baltica* to 50 µg of heptachlor L^{-1} reduced cell density, ^{14}C uptake (as $[^{14}C]O_2$) per cell, chlorophyll *a* levels per unit culture volume, and carbon fixation per unit chlorophyll *a* (Magnani et al. 1978). Pardini et al. (1971) suggested that heptachlor might interfere with electron transport in the dinoflagellate, as it does in mammalian cells. Dinoflagellates, like other protozoans, are at the base of the food web, so changes in normal function and growth caused by pesticides could influence food availability in aquatic food webs.

Mediterranean mussels (*Mytilus galloprovincialis*) and deep-water pink shrimp (*Parapenaeus longirostris*) from four sections of the Saronikos Gulf, Greece, were analyzed for heptachlor epoxide residues (Satsmadjis and Voutsinou-Taliadouri 1983). Mean burdens of heptachlor epoxide were 0.18 and 0.01 µg g^{-1} (lw) in mussels and shrimp, respectively. No heptachlor or heptachlor epoxide residues were detected in translocated mussels or in water samples taken from a site near the ocean outfall for the Los Angeles County sewer system at White Point, California (Green et al. 1986). Salinity also influences heptachlor toxicity: grass shrimp (*Palaemonetes vulgaris*) were less affected by heptachlor in 48-hr tests at a salinity of 18 $^o/_{oo}$ (parts per thousand) than at 12 $^o/_{oo}$ (Eisler 1969).

Acute toxicity and residue data for heptachlor in fish are extremely varied. Certainly, the various locations of sampling and inputs of heptachlor at those locations will affect residue levels; however, one should also consider the different detection limits of the analytical procedures. In very few cases did authors report detection limits, making comparisons of data very difficult.

Al-Omar et al. (1986) examined 11 fish species from a polluted region in the Diyala River (Baghdad, Iraq) for heptachlor and its epoxide. Residues ranged from 0.333 to 4.012 µg g^{-1} (lw). Even though heptachlor per se has never been

Table 4. Summary of heptachlor LC$_{50}$s to aquatic organisms

Species	Exposure µg L^{-1}	hr	Reference
Phytoplankton	1 (94%)[a]	4	Butler 1963
Insecta			
stonefly	0.9–1.1	96	Sanders and Cope 1968
Pteronarcella sp.	0.9	96	Johnson and Finley 1980
Pteronarcys sp.	1.1	96	Johnson and Finley 1980
Claassenia sp.	2.8	96	Johnson and Finley 1980
Crustacea			
Gammarus lacustris	150	24	Sanders 1969
hermit crab	470	24	Eisler 1969
Daphnia pulex	42[b]	48	Sanders and Cope 1966
Daphnia magna	78	48	Macek et al. 1976
	57.7	50	Anderson 1971
Orconectes sp.	0.5	96	Johnson and Finley 1980
pink shrimp	0.11	96	Schimmel et al. 1976
Palaemonetes sp.	1.8	96	Johnson and Finley 1980
Gammarus lacustris	29	96	Johnson and Finley 1980
Simocephalus sp.	47	96	Johnson and Finley 1980
hermit crab	55	96	Eisler 1969
Gammarus fasciatus	56	96	Johnson and Finley 1980
Pelecypoda			
American oyster	1.5	96	Schimmel et al. 1976
Osteichthyes			
Fundulus similis	4.8	24	Holden 1973
	3	48	Holden 1973
Menidia menidia	3.0	96	Eisler 1970
rainbow trout	7.0	96	Macek et al. 1969
	7.4	96	Johnson and Finley 1980
	20	96	Johnson and Finley 1980
American eel	10	96	Eisler 1970
bluegill	13	96	Johnson and Finley 1980
	26	96	Henderson et al. 1969
redear sunfish	17	96	Johnson and Finley 1980
fathead minnow	23	96	Johnson and Finley 1980
	78	96	Henderson et al. 1969
	130	96	Henderson et al. 1969
channel catfish	25	96	Johnson and Finley 1980
Fundulus heteroclitus	50	96	Eisler 1970

[a] % reduction in carbon fixation in a mixed culture of marine phytoplankton.
[b] Concentration required to immobilize 50% of the organisms at 15.6°C.

used in Iraq, residues were found in two fish species from the Shatt al-Arab River (Dou Abul et al. 1987). Total heptachlor plus heptachlor epoxide residues from a cyprinid (*Barbus xanthopterus*) and the Indian shad (*Tenualosa ilisha*) collected from the river averaged 0.003 µg g^{-1} and 0.017 µg g^{-1} (ww), respectively. Concentrations of 2-91 ng kg^{-1} (sic) in water, 48-204 ng kg^{-1} on suspended particulate matter, and 13-48 ng kg^{-1} in sediment were also found in the river (Dou Abul et al. 1988). The detection of residues may be due to the presence of heptachlor (10-11%) in chlordane, which is applied in Iraq (Braun and Frank 1980). Cutthroat trout (*Salmo clarki*) from a lake that had been sprayed with technical chlordane had detectable body residues of both heptachlor and heptachlor epoxide (Albright et al. 1980). Trout from four rivers in León, Spain, were analyzed for heptachlor epoxide residues (Teran and Sierra 1987). Liver, brain, kidney, and muscle were assayed for the epoxide with mean concentrations of 0.016, 0.035, 0.018, and 0.007 µg g^{-1} (ww), respectively. Heptachlor residues were found in 21-78% of muscle tissue of six fish species from four rivers in southern Italy with concentrations ranging from 5 to 20 ng g^{-1}, while heptachlor epoxide was found in 100% of the samples at 5-16 ng g^{-1} (Amodio-Coccieri and Arnese 1988). Spottail shiners from Lakes Ontario, Erie, and St. Clair were found to have body burdens of 1-2 ng heptachlor g^{-1} (ww) (Suns and Rees 1978). Frank et al. (1978) analyzed fish from Lake Simcoe, Ontario, for organochlorine residues. Heptachlor epoxide concentrations in smaller male and female fish (<4 kg) were, respectively, 0.003 and 0.002 µg g^{-1} (ww) in tissues, and 0.06 and 0.03 µg g^{-1} in fat. In larger fish (>4 kg), the heptachlor epoxide residues increased significantly: 0.043 and 0.027 µg g^{-1} (ww) in tissues, and 0.40 and 0.30 µg g^{-1} (ww) in fat, for males and females respectively.

Leung et al. (1981) measured whole-body burden of heptachlor epoxide in seven fish species from a reservoir in Iowa. Heptachlor epoxide ranged from nd-0.07 µg g^{-1} (ww), with the mean of approximately 0.001 µg g^{-1} (ww). In the Des Moines River, adjacent to the reservoir, the same seven fish species had levels of the epoxide similar to those in the above study (Bulkley et al. 1981). The authors also found that heptachlor epoxide accumulated to a greater extent in fish with higher fat content, such as gizzard shad and channel catfish. In Tuttle Creek Lake, Kansas, heptachlor epoxide was detected at concentrations of 0.008 and 0.005 mg kg^{-1} in common carp (*Cyprinus carpio*) and white bass (*Morone chrysops*), respectively (Arruda et al. 1988). In 1980-1981, as part of the National Pesticide Monitoring Program, the US Fish and Wildlife Service sponsored a nationwide collection of fish for assessment of whole-body organochlorine pesticide content (Schmitt et al. 1985). Since heptachlor is rapidly metabolized in vivo, no parent compound was detected in any of the fish samples. Heptachlor epoxide was present in many of the fish (0.27 µg g^{-1} (ww), maximum residue); however, both the incidence and magnitude of heptachlor residues decreased when compared to 1978-1979 samples due to a decrease in the agricultural use of heptachlor.

Ober et al. (1987) sampled 11 marine fish and shellfish species from the Chilean coastal waters and found heptachlor in only species, *Seriolella violacea* (amberjack) at 0.9 µg g^{-1} (ww). Salmon captured in the Caspian Sea had heptachlor residues of 0.053 µg g^{-1}; however, none of the other species sampled contained any heptachlor (Hashemy-Tonkabony and Langaroodi 1976). Heptachlor residues were present in winter flounder (*Pseudopleuronectes americanus*) tissues from the Weweantic River estuary, Massachusetts (Smith and Cole 1970). In juvenile fish, levels ranged from 0.01 to 1.10 and 0.01 to 0.44 µg g^{-1} (ww) for heptachlor and heptachlor epoxide, respectively. Striped mullet (*Mullus barbatus*) from the Saronikos Gulf, Greece, contained nd-0.003 µg heptachlor epoxide g^{-1} as measured in lyophilized fleshy parts of fish (Voutsinou-Taliadouri and Satsmadjis 1982). Heptachlor residues were not detected in the livers of *Antimora rostrata*, a bathydemersal fish (Barber and Warlen 1979).

Harp seal tissue samples from five locations in the northwest Atlantic and Arctic Oceans were analyzed for organochlorine residues (Ronald et al. 1984). Blubber held the greatest concentration of heptachlor epoxide, and adult males had the highest body burdens [0.07 µg g^{-1} (ww) mean residue]. Residue levels ranged from 0.01 to 0.19 µg g^{-1} (ww), varying with location of capture, age, and sex. Harp seals are opportunistic feeders, eating pelagic fish and shellfish, and likely receive their pesticide body burden by ingestion.

In laboratory studies, heptachlor persisted in fish for approximately 1 mon (Macek 1970). Goldfish eliminated 18% of intraperitoneally injected heptachlor in 10 d (Khan et al. 1979). The polar metabolites recovered were heptachlor, heptachlor-2,3-epoxide, 1-hydroxychlordene, and 1-hydroxy-2,3-epoxychlordene. Shaffi (1979) exposed nine species of tropical fish to 1–5 µg of heptachlor L^{-1}. All fish experienced acute respiratory distress with mucus accumulation in the gill cavity. In general, there was an inverse relationship between tissue glycogen content and heptachlor dose, while there was a direct correlation of glucose and serum lactate accumulation with dose. Heptachlor appears to act by damaging gill structures, thus forcing the fish to obtain needed energy by anaerobic metabolism or glycogenolysis. Radhaiah et al. (1988) exposed *Tilapia mossambica* to 0.03 mg heptachlor L^{-1} (20% of the LC$_{50}$) for 5, 10, and 15 d. They found decreased carbohydrate content in kidneys of exposed fish with increased activities of lactate and glutamate dehydrogenases. These changes signal an increased energy demand undoubtedly to overcome the pesticide stress.

The microsomal fractions of fish brain and liver homogenates are sensitive to heptachlor exposure. Exposure to 37.35 mg heptachlor L^{-1} caused a 67% inhibition of Na$^+$, K$^+$-ATPases, and a 70% inhibition of Mg^{2+}-ATPase in rainbow trout gill microsomes (Davis et al. 1972). Similarly, using bluegill, Yap et al. (1975) found 6.8 mg heptachlor L^{-1} added to the incubation medium produced 50% inhibition of liver mitochondrial ATPase, and exposure to 16.4 mg L^{-1} caused 50% inhibition of Na$^+$, K$^+$-ATPase in brain homogenate. Bluegill brain Na$^+$, K$^+$-ATPases and Mg^{2+}-ATPase were inhibited 58.6% and 65.6%, respectively, by

15.6 mg L^{-1} heptachlor in the culture medium (Cutkomp et al. 1971). Both oxygen and phosphate utilization by bluegill liver mitochondria were reduced by incorporation of 37 mg heptachlor L^{-1} in the incubation medium (Hiltibran 1974).

Heptachlor toxicity in fish may be modified by temperature. Bridges (1965) showed a reduction in mortality with increased temperature in redear sunfish (*Lepomis microlophus*) exposed to heptachlor. Conversely, Macek et al. (1969) found rainbow trout to be more susceptible to heptachlor toxicity at higher temperatures during 24- and 96-hr LC$_{50}$ assays. The warm-water sunfish may alter behavior (i.e., become less active) in response to temperature increases above "normal," whereas a cold-water fish such as the trout may increase its metabolic rate in response to increased temperature, thereby increasing the rate of heptachlor metabolism.

C. Bioaccumulation

The hydrophobic nature of organochlorine compounds makes them resistant to common transformation and detoxication pathways, resulting in extensive binding of the pesticide to various lipid-type compartments in the exposed system. This binding enhances the environmental stability, persistence, and accessibility of heptachlor to organisms (Kerr and Vass 1973). Pesticide residue levels may gradually increase over time via food-web transfers, ultimately producing harmful body concentrations.

The bioconcentration factor (BCF) of a compound is the ratio of the concentration in the organism to the concentration in the diet, and is an indicator of the organism's tendency to accumulate a compound above input concentration. For domestic species, dogs fed a diet containing heptachlor for 12 wk showed a BCF of 4.1–6.2 for females and 0.5–1.4 for males (Radomski and Davidow 1953), while chickens fed heptachlor showed a BCF of 6.0 (Kan and Tuinstra 1976). As an index of bioaccumulation potential in nondomestic animals, the ratio of storage to exposure was 5.5:1, for the geometric mean concentration of heptachlor epoxide in earthworms sampled from heptachlor treated soils (Gish 1970).

In the case of heptachlor, there is some controversy over its bioaccumulation potential. Normally, adipose tissue is considered an essentially limitless storage depot for lipophilic compounds, and the question of steady-state concentration does not arise. However, with heptachlor, experimental evidence indicated that steady-state levels are in fact reached after a period of time. Broiler chickens exposed to 0.3 mg of heptachlor kg^{-1} for 8 wk reached plateau residue levels (primarily as heptachlor epoxide) in adipose tissue after 4 wk (Wagstaff et al. 1980). This study, however, may not represent a true picture of residue accumulation in fat because it was conducted during the first weeks of the life of the chicks. During that time, body fat in chicks is redistributed, which may mobilize stored residues and make them more readily eliminated.

In terrestrial systems, heptachlor can transfer from soil to both edible and nonedible foliage in plants (residue amounts are detailed in section V.A), and ultimately to animal species feeding upon contaminated plants. Other animals accumulate heptachlor residues directly, e.g., soil inhabiting invertebrates. Jeffries and Prestt (1966) stated that lethal residues in carnivorous animals may result from ingestion of a few highly contaminated prey animals, rather than chronic low level exposure. A similar conclusion was reached by DeWeese et al. (1986), who suggested that potentially harmful organochlorine concentrations are present in certain western migratory birds, and pose an even greater hazard to avian predators such as the peregrine (due to bioaccumulation). Burrage and Saha (1972) proposed that heptachlor treatment of grain seed be discontinued in favor of lindane seed dressing, due to rapid accumulation of heptachlor residues in tissues of seed-eating birds, further supporting the conclusions of DeWeese et al. (1986).

To assess food chain transfers in an aquatic ecosystem, Norstrom et al. (1978) analyzed coho salmon, herring gulls, and alewives in Lake Ontario for organochlorine contamination, as both salmon and herring gulls depend on alewives and other small fish for food. Heptachlor epoxide residues were 0.003 $\mu g\ g^{-1}$ (ww) in alewives, 0.015 and 0.007 $\mu g\ g^{-1}$ (ww) in salmon muscle and liver, respectively, and 0.1–0.15 $\mu g\ g^{-1}$ (ww) in herring gull eggs, indicating a degree of bioaccumulation along the food web.

In addition to higher order mammalian and avian predators, humans may serve as potential endpoints of ecosystem transfers of heptachlor. Heptachlor applied at 2.25 kg ha^{-1} to a Louisiana soil, produced earthworm residues often >3 μg heptachlor epoxide g^{-1}; 10 of 12 woodcock feeding on these earthworms died within 55 d (Stickel et al. 1965a). Eleven of 12 birds survived 60 d on a diet of earthworms containing average residue levels of 0.65 $\mu g\ g^{-1}$ (the lone bird died from apparently unrelated causes) (Stickel et al. 1965a). The response of woodcock in these feeding studies is closely related to the physical condition of the birds, especially body weight (Stickel et al. 1965b). Beyer and Gish (1980) found similar results, and suggested that agricultural plots, treated with a "normal" application rate of heptachlor (2.2 kg ha^{-1}), remained hazardous to woodcock (via residue transfer through earthworms) for approximately 3 yr, with longer hazardous periods for greater application rates. There may be a threat to humans who regularly ate gamebirds, as muscle samples of grouse, quail, and woodcock contained >1 μg of heptachlor epoxide g^{-1} (ww), exceeding the WHO residue tolerance limits of 0.03–0.3 μg heptachlor g^{-1} food d^{-1} (Blevins 1979).

Another potential route of exposure for humans is through mammalian milk. Residues levels in the milk of cows fed hay contaminated with 1 μg of heptachlor g^{-1} produced milk which contained 1.94 μg of heptachlor epoxide g^{-1} butterfat (Bache et al. 1960). In Australia, McDougall et al. (1987) reported that soil levels of heptachlor and heptachlor epoxide approaching 0.1 $\mu g\ g^{-1}$ can result in milk residue levels that exceed their legislated standard (0.15 $\mu g\ kg^{-1}$ in butterfat) when cows are allowed to graze on treated land. Thus, the reported levels of heptachlor and heptachlor epoxide in agricultural soils would warrant closer scrutiny if field use of the pesticide continued. Human milk may also serve as a conduit for exposure of infants to pesticide residues (section XI.B).

VI. Physiological Properties of Heptachlor

Complete information is lacking with respect to the metabolic fate of heptachlor in animals and humans. As a group, the chlorinated hydrocarbon insecticides have received considerable study, and some of the cyclodiene insecticides (dieldrin in particular) have been studied in detail (Geyer et al. 1986). However, individual compounds within a class of insecticides may behave differently from each other in terms of metabolism and pharmacokinetics, and each insecticide should therefore be studied independently.

Due to inherent difficulty in sampling the appropriate biological reservoir (adipose tissue), detection of heptachlor in humans has been limited to necropsy samples of adults and neonates, in vitro microsomal preparations of human liver, and milk of nursing mothers. Additional metabolic information is gathered from in vivo and in vitro studies on laboratory and domestic organisms.

A. Metabolism

Heptachlor epoxide, the primary metabolite of heptachlor, was first identified in dogs administered heptachlor in their diet (Davidow and Radomski 1953). Epoxidation of the parent compound appears to be the most important metabolic reaction, with other derivatives forming as oxidized products and conjugates. The most complete study available examines the metabolites found in the feces of rats given oral doses of [^{14}C]heptachlor (Tashiro and Matsumura 1978). Extensive characterization of fecal matter revealed the excreta to contain 26.2% heptachlor, 19.5% 1-hydroxychlordene, 17.5% 1-hydroxy-2,3-epoxychlordene, 13.1% heptachlor epoxide, 3.5% 1,2-dihydroxychlordene, and 19.0% an unidentified metabolite. An in vitro study by the same group with rat liver microsomes indicated that a much higher proportion of the metabolites was heptachlor epoxide (85.8%) and the other metabolites accounted for <5% each.

In contrast, sheep metabolize heptachlor rapidly to primarily water-soluble residues which are excreted in the urine (Holcombe et al. 1987; Smith et al. 1987a,b). Specifically, sheep eliminated 34% of a 1.64 mg kg^{-1} dose in 21 d, ⅔ in the urine and ⅓ in the feces (Smith et al. 1987b). While secretion of lipophilic metabolites into the milk is a significant pathway in some mammalian species (cf. section XI.B), the milk is not a significant depot for lactating sheep. Lambs nursing following administration of 2 mg kg^{-1} [^{14}C]heptachlor to their mothers did not receive a detectable dose of heptachlor or heptachlor metabolites (Smith et al. 1987a).

Essentially the same metabolites as those found in rats have been detected using human liver microsomes, although the distribution is quite different (Tashiro and Matsumura 1978): 68.6% unmetabolized heptachlor, 20.4% heptachlor epoxide, 5.0% 1-hydroxy-2,3-epoxychlordene, 4.8% 1-hydroxychlordene, 0.1% 1,2-dihydroxydihydrochlordene, and 1.0% unidentified conjugate. Thus, in humans a much smaller proportion is converted to heptachlor epoxide.

Possible pathways for the metabolism of heptachlor are shown in Fig. 3. Heptachlor epoxide has also been shown to act as a substrate for enzymatic oxidation in both pig and sheep microsomal systems (Brooks 1969), forming dihydroxydihydrochlordene as the major product. Another study demonstrated that heptachlor undergoes reductive metabolism in vitro (Yoneyama and Matsumura 1981). Under anaerobic conditions using rat liver microsomes and flavin cofactors, heptachlor was converted to chlordene by a reducing dechlorination reaction. The significance in vivo of this finding was not well defined.

The enzyme that carries out the epoxidation reaction was first termed epoxidase, but more recent evidence suggests that the reaction is mediated by the mixed function oxidase (MFO) system of the liver (Greene 1972). In microsomal systems, the oxidation of heptachlor requires the presence of both NADPH and O_2. Heptachlor binds to the enzyme with high affinity, and produces a Type I difference spectrum. Other cyclodiene insecticides show similar activity in the MFO system.

Attempts have been made to characterize Phase II (conjugation) products of heptachlor metabolism. Goldfish injected with [^{14}C]heptachlor excreted a conjugated compound in urine and feces, but analysis yielded unsatisfactory results (Feroz and Khan 1979). Treatment of this conjugate with β-glucuronidase did not release heptachlor, indicating it is not a glucuronidated product; acid hydrolysis produced a trihydroxylated derivative, but there was no analysis of the other cleavage products. Heptachlor epoxide is a good substrate for glutathione-S-transferase, but the actual in vivo conjugate has not been identified (Scheufler and Rozman 1984). Glutathione conjugation is an important mechanism for the elimination of epoxides in mammals.

There are a wide variety of conditions and treatments which alter heptachlor metabolism. Among the most interesting and promising as poisoning treatment protocols are studies involving the reduction of heptachlor body burdens in rats by induction of the Phase II enzymes (Rozman 1984; Scheufler and Rozman 1984). Hexadecane enhances transfer of lipophilic compounds into the lumen of the intestine, while *trans*-stilbene oxide (TSO) is both an enhancer of biliary excretion and an inducer of epoxide hydrolases and glutathione-S-transferases. Hexadecane administered 24 hr before heptachlor dosing had little effect on heptachlor's half-life over a 30-d period, while TSO treatment 17–20 d after heptachlor dosing in hexadecane-treated animals shortened the half-life threefold. Thus, Phase II enzyme inducers may be effective in increasing the elimination of this pesticide.

Dietary lipids affect the ability to metabolize pesticides (Caster et al. 1970; Wade and Norred 1976). Using liver microsomes prepared from rats fed varied concentrations of linoleate, the authors found higher levels of lipid intake (up to 10% of calories) significantly increased epoxidase activity. They postulated that this increase in activity was due to increases in the microsomal P-450 enzymes. In another study, the influences of protein in the diet were examined (Weatherholtz and Webb 1971). Rats fed protein-deficient diets of 5% casein (vs 20% in controls), showed 50% inhibition of microsomal epoxidase activity relative to controls. Interestingly, the reduction in epoxidase activity correlated with the increasing toxicity of the compound to weanling rats. This suggested that heptachlor epoxide is the compound responsible for the toxic effects.

B. Absorption, Distribution, and Elimination

Heptachlor is highly lipophilic and readily absorbed by most routes of exposure. In the past, human exposure to heptachlor most likely resulted from food contamination; as such, gastrointestinal absorption was the commonly used route when studying the quantitative kinetics of heptachlor distribution. An acute intravenous dose of heptachlor administered to rats resulted in systemic distribution of the compound, and high levels of heptachlor were present in those areas more heavily perfused, especially liver, kidney, and to some extent, skin (Scheufler and Rozman 1984). Chronic exposure results in a somewhat different profile of tissue heptachlor levels. As expected, heptachlor (primarily as the epoxide) accumulates readily in adipose tissue and other tissues with some fat content, including the liver and kidney.

A number of studies have assessed distribution through the placenta into the tissues of stillborn and newborn babies. Levels of heptachlor epoxide averaging $0.173 \, \mu g \, g^{-1}$ ($0.191 \, \mu g \, g^{-1}$ in males, $0.141 \, \mu g \, g^{-1}$ in females) were detected in a study of 52 neonates from 13 US cities (Zavon et al. 1969), indicating transplacental exposure to the pesticide. Tissue distribution of heptachlor epoxide in another study on stillborn babies showed residues were present in heart, liver, kidney, adrenal glands, and adipose tissue (Curley et al. 1969).

Elimination of heptachlor from the body occurs primarily via fecal and urinary excretion, with the fecal route being the most important (approximately 90% vs 10% urinary) (Scheufler and Rozman 1984). Another important route is through the milk of lactating animals. In cows fed hay containing various levels of heptachlor over a period of 45 d, heptachlor epoxide did not appear in the milk until the tenth day, indicating that transfer of the compound to this compartment is a relatively slow process (Huber and Bishop 1961). Elimination of heptachlor followed first-order kinetics, that is, the rate of excretion depends directly on the amount of the compound in the body. Estimates of the half-life of heptachlor vary considerably between different species. Chronic feeding studies produced half-lives of approximately 4 wk in chickens (Wagstaff et al. 1980) and 3 wk in dogs

Table 5. Microsomal enzyme induction by heptachlor

Isozyme activity	Other effects	Reference
Aldrin epoxidase	increased liver weight	Campbell et al. 1983 Haake et al. 1987
Aminopyrine demethylase		Den Tonkelaar and Van Esch 1974
Aniline hydroxylase		Den Tonkelaar and Van Esch 1974
Benzo[a]pyrene hydroxylase		Haake et al. 1987
Cytochrome b_5	increased liver weight	Campbell et al. 1983
Dimethylaminoantipyrine	increased liver weight	Campbell et al. 1983, Haake et al. 1987
Ethoxyresorufin-O-deethylase		Haake et al. 1987
Hexobarbital oxidase		Den Tonkelaar and Van Esch 1974
N-demethylase	increased liver weight	Campbell et al. 1983, Kinoshita and Kempf 1970
O-demethylase		Kinoshita and Kempf 1970
Phosphorothioate detoxication		Kinoshita and Kempf 1970
Testosterone 16α- and 16β-hydrolases		Haake et al. 1987

(Radomski and Davidow 1953). Acute (single dose) studies show much shorter half-lives, probably due to limited storage in fatty tissues: 5 d in rats (Rozman 1984), 2–4 d in rats (Scheufler and Rozman 1984), and 28 d in goldfish (Feroz and Khan 1979).

In an attempt to correlate heptachlor concentration in bat guano with carcass concentration, Clark et al. (1981) found this method unreliable for the prediction of residue levels in the body. In cases of contamination by heptachlor where the exposure level is unknown, one should be very careful in making predictions of total body burdens of heptachlor based upon analysis of excreted material.

VII. Organ System Toxicity

The effects of heptachlor on the liver have been studied extensively. Heptachlor stimulated NADPH oxidase in male rat microsomes in vitro, and increased the rate of cytochrome P-450 reduction by 150–200%. The affinity of heptachlor for cytochrome P-450 and stimulation of NADPH oxidase was markedly less in female rat microsomes, and the rate of reduction of cytochrome P-450 was unchanged by the presence of heptachlor (Greene 1972). Following the observation that heptachlor and heptachlor epoxide decreased hexobarbital sleeping times in male rats (Conney et al. 1967), numerous enzyme activities induced by heptachlor were described, and are summarized in Table 5. Although Den

Tonkelaar and Van Esch (1974) established a no observable effect level (NOEL) of approximately 1 µg g^{-1} for microsomal enzyme induction by heptachlor, heptachlor induced P-450b and P-450d (the phenobarbital (PB)-inducible isozymes), as well as other P-450 isozymes at a dose of 250 µmol kg^{-1} (93 mg kg^{-1}). These other inducible isozymes were responsible for differences in testosterone metabolites seen following PB induction and heptachlor treatment (Haake et al. 1987).

While enzyme induction is not necessarily a measure or indication of toxicity, increased P-450 activity might be an exposure marker. However, enzyme induction may have unwanted or even deleterious effects in cases where the desired effects of therapeutic agents are diminished or the toxicity of xenobiotics exacerbated by enhanced metabolism. For example, enzyme induction by heptachlor caused an increase in the ED_{50} for the antipyretic/analgesic aminopyrine (Vargova and Kovalick 1969). Induction of P-450 enzymes can both accelerate the rate at which procarcinogens are activated and decrease the efficacy of chemotherapeutic agents (Guengerich 1988).

Following administration of 2-5 mg of heptachlor kg^{-1}, pig liver cells showed ultrastructural changes including glycogen depletion, increased amounts of smooth endoplasmic reticulum (SER), and widening of the cisternae of the rough endoplasmic reticulum (Dvorak and Halacka 1975). There is a relationship between potent liver enzyme inducers, hepatocyte SER proliferation (thought to be the site of origin of the enzyme production), and increased metabolic activity of the thyroid. Effects on the thyroid include elevation of TSH and depletion of T_4, which some feel may be removed from the circulation by the proliferated hepatic SER (Bernstein et al. 1968; Lissitsky 1976; Lumb and Rust 1985). An increased uptake of ^{125}I-T_4 by the liver is associated with microsomal enzyme induction and SER proliferation in rats exposed to chlordane (Bernstein et al. 1968). These observations taken together are of interest, as heptachlor was shown to be associated with an increased number of follicular cell carcinomas in the NCI (1977b) heptachlor carcinogenicity bioassay (cf. section IX).

Overtly detrimental effects of chronic heptachlor on the liver were seen in mice receiving 10 mg kg^{-1} heptachlor or heptachlor epoxide in their diet (Reuber 1977a). Animals had a high incidence of hepatic vein thrombosis (HVT), frequently accompanied by infarcts, cirrhosis, carcinomas, and/or thrombi in the atria of the heart. Female mice were more susceptible to HVT than male mice.

Heptachlor had multiple, concentration-dependent actions on oxidative phosphorylation in rat liver mitochondria. State 3 respiration was inhibited 50% by 20-25 nanomol (nmol) heptachlor mg^{-1} of mitochondrial protein (Nelson 1975). The onset of uncoupling occurred near 20 nmol mg^{-1}, and maximal activation of State 4 respiration was achieved by an additional 10-15 nmol mg^{-1}. Direct inhibition of electron transport processes occurred at 200-500 nmol heptachlor mg^{-1} protein. Gasper and Kawatski (1972) demonstrated that succinic dehydrogenase activity was slightly inhibited in mouse liver homogenates by 1.66×10^{-4} M heptachlor. The biological relevance of these data is in question, as this concentration is more than an order of magnitude greater than the highest levels reported by Barquet et al. (1981) in human fat, or in human milk following confirmed environmental exposure (Takahashi et al. 1981).

Heptachlor altered normal glucose homeostasis in rats. Either a single dose or chronic exposure to 15 mg heptachlor kg^{-1} stimulated hepatic pyruvate carboxylase, phosphoenolpyruvate carboxykinase, fructose 1,6-bisphosphate phosphatase, and glucose-6-phosphate phosphatase. Correspondingly, liver glycogen levels fell, accompanied by a rise in blood glucose and urea. These effects were attributed to a stimulatory effect by heptachlor on the cyclic AMP-adenlyate cyclase system (Singhal and Kacew 1976).

Heptachlor hepatocarcinogenicity will be discussed in section IX.

Seizures and evidence of cortical excitability are the primary central nervous system symptoms following acute heptachlor exposure. Joy (1976) demonstrated that seizure activity began in cats 20–40 min after intravenous administration of 2–10 mg heptachlor kg^{-1}. While heptachlor produced enhanced cortical responses to sensory stimuli, primary responses along the somatosensory pathways in the medial lemniscus and the ventral posterior lateral nucleus of the thalamus were depressed. Similarly, St. Omer and Ecobichon (1971) noted increased motor activity in rats shortly after intracarotid administration of heptachlor. Periods of increased motor activity were followed by hypokinesia with dyspnea, mild to severe tremors, and seizures. Although seizures have been attributed to increased neurotransmitter release, acetylcholine (ACh) levels were elevated 2.3-fold only during the period of increased motor activity, and decreased and returned to normal during the most severe seizure activity. Further studies (Hrdina et al. 1974) showed that chronic heptachlor (3 or 15 mg kg^{-1}) or heptachlor epoxide (1 or 5 mg kg^{-1}) administration significantly decreased cerebrocortical ACh, increased levels of brainstem serotonin, and produced no change in brainstem norepinephrine levels with no overt signs of neurotoxicity.

A number of studies have attempted to explain the cortical excitability and altered neurotransmitter levels associated with heptachlor exposure. Two theories have received a great deal of attention. The first hypothesizes that heptachlor inhibits a Ca^{2+}, Mg^{2+}-ATPase, causing elevated intracellular Ca^{2+} levels with subsequent increased release of excitatory neurotransmitters. Both sequestration of Ca^{2+} in the endoplasmic reticulum and extrusion of Ca^{2+} from the cell are Ca^{2+}, Mg^{2+}-ATPase-dependent processes. Heptachlor stimulated the release of the excitatory neurotransmitter glutamate from rat brain synaptosomes (Yamaguchi et al. 1980); glutamate release is also a Ca^{2+}-dependent process (Puszkin and Kochwa 1974). Heptachlor epoxide-treated synaptosomes took up more and released less Ca^{2+} than untreated synaptosomes (Yamaguchi et al. 1980). Synaptosomes isolated from heptachlor-treated rats showed the same trends toward increased neurotransmitter release and altered Ca^{2+} homeostasis as were seen in vitro (Yamaguchi et al. 1979, 1980). Increased intracellular Ca^{2+} levels, then, are associated with a net increase in neurotransmitter release. Fig. 4 summarizes this hypothesis.

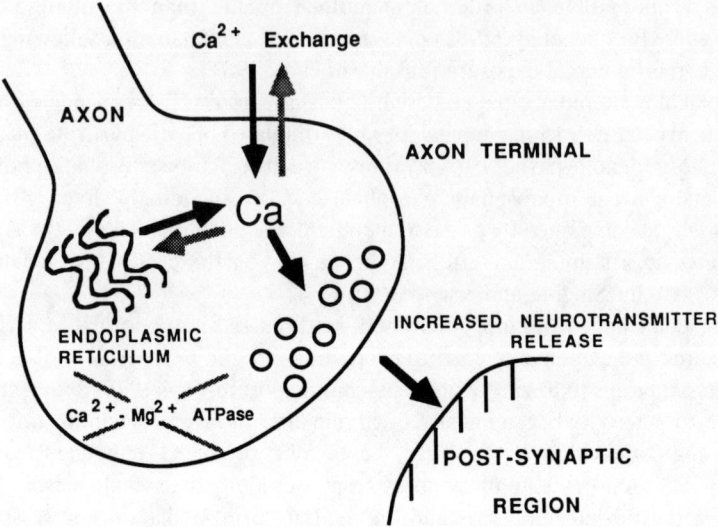

Fig. 4. Schematic illustration of the effects of heptachlor epoxide on neurotransmitter release. Solid arrows represent stimulation by heptachlor epoxide, shaded arrows show inhibition. The overall effect of heptachlor epoxide is the increase of internal Ca^{2+}, which triggers neurotransmitter release. Reprinted with permission from Pergamon Journals, Inc.: Yamaguchi I, Matsumura F, Kadous AA (1980) Heptachlor epoxide: Effects on calcium-mediated transmitter release from brain synaptosomes in rat. Biochem Pharmacol 29:1815–1824.

A second theory to explain the CNS effects of heptachlor involve the γ-aminobutyric acid (GABA) binding sites on postsynaptic membranes. The major inhibitory neurotransmitter in the central nervous system (and at neuromuscular junctions in lower animals) is GABA (White et al. 1978). The GABA receptor is part of a complex containing a GABA site and a Cl⁻ ionophore (Matsumura and Tanaka 1984) with binding capacities for benzodiazepines and picrotoxin (Abalis et al. 1985). Activation of GABA receptors increased membrane conductance for Cl⁻ in postsynaptic membranes, such that when an excitatory impulse is delivered, e.g., by glutamate or ACh, the postsynaptic membrane does not reach threshold, and transmission of the impulse does not occur (White et al. 1978). [^{35}S]t-butylbicyclophosphorothionate ([^{35}S]TBPS) is a highly specific noncompetitive allosteric inhibitor of GABA which binds to the Cl⁻ ion channel of the GABA complex (Squires et al. 1983) and is used as a label for examining binding to the GABA receptor site. Both heptachlor and heptachlor epoxide were effective in displacing [^{35}S]TBPS from GABA binding sites in rat brain membrane preparations, with IC_{50}'s of 400 nM and 70 nM, respectively (Abalis et al. 1985). A good correlation was shown between the efficiency of heptachlor in inhibiting GABA, its potency in inhibiting GABA-induced $^{36}Cl^-$ influx, and its ability to inhibit the binding of [^{35}S]TBPS to rat brain microsacs. These results were interpreted to mean that the GABA receptor in the mammalian brain is a primary target for the toxic action of heptachlor (Abalis et al. 1985; Gant et al. 1987). Alternatively, Ghiasuddin and Matsumura (1982) found that the heptachlor epoxide bound avidly to picrotoxin receptors, suggesting that it is the Cl$^-$ ionophore which is the target of heptachlor epoxide. Interestingly, the picrotoxin receptor of the American cockroach was much more sensitive to heptachlor epoxide than that of the rat brain (Matsumura and Ghiasuddin 1982; Matsumura and Tanaka 1984).

Heptachlor has a number of effects on reproduction in experimental animals. A dose of 10 mg of heptachlor kg^{-1} increased metabolism of estrone by rat liver microsomes. The action of endogenous estrogens on the uterus was diminished, and decreased action of exogenously administered estrone was noted (Welch et al. 1971). Heptachlor, at a concentration of 7 mg L^{-1}, caused abnormal development and lysis of two-cell sea urchin embryos. Fertilization was also markedly decreased, but with no apparent effect on sperm motility (Bresch and Arendt 1977). In male rats, 20 mg of heptachlor kg^{-1} in the diet did not inhibit 5α- dihydroxytestosterone binding to prostatic receptors, but subchronic (90 d) administration of heptachlor decreased the androgen receptor content of the rat ventral prostate (Shain et al. 1977).

Several components of the immune system are affected by heptachlor exposure. Heptachlor can cause contact dermatitis in humans (Gosselin et al. 1976). Heptachlor and heptachlor epoxide (0.01–16 mg L^{-1}) stimulated dose-dependent histamine release from rat peritoneal mast cells and human basophils in vitro (Rohr et al. 1985). Heptachlor and heptachlor epoxide also stimulated mast cells to release the leukotrienes C3 and C4, which are chemotactic for eosinophils and neutrophils. In vivo studies in 3 to 8-week-old chicks by Rodica and Stefania (1973) showed that 1 μg of heptachlor g^{-1} in the diet caused a 28% increase in the weight of the bursa of Fabricius, the site of maturation of antibody-forming B-cells.

In the kidney, heptachlor as a single dose (200 mg kg^{-1}) or chronically administered (15 mg kg^{-1} for 45 d) stimulated the activities of pyruvate carboxylase, phosphoenolpyruvate carboxykinase, fructose 1,6-bisphosphate phosphatase, and glucose 6-phosphate phosphatase. These alterations in glucose homeostasis were attributed to stimulation of the cyclic AMP-adenylate cyclase systems in the kidney (and liver) cortex (Singhal and Kacew 1976). The presence of 10^{-6} M heptachlor or heptachlor epoxide increased the amount of cyclic [^{3}H]AMP synthesized from tritiated adenosine in renal cortical slices by about 150% (Kacew and Singhal 1974).

Heptachlor inhibited heme synthesis in chick embryos (10 or 20 mg injected per egg) and in 1-day old chickens (10 or 20 mg intraperitoneally) by blocking uroporphyrinogen decarboxylase, which converts uroporphyrinogen III to coproporphyrinogen III (Taira and San Martin de Viale 1980). Inhibition of heme synthesis by heptachlor was reduced by SKF525-A, an inhibitor of cytochrome P-450, and enhanced by phenobarbital, an inducer of cytochrome P-450; thus, a metabolite of heptachlor appears responsible for blocking heme synthesis. Inhibition of heme synthesis might depress levels of hemoglobin, catalase, and all cytochromes, potentially causing microcytic anemia.

Heptachlor alters ion transport mechanisms. In vitro studies by Webb et al. (1976) showed that 10^{-5} M heptachlor inhibited Na$^+$, K$^+$-ATPase in mature human erythrocytes. The clinical relevance of this treatment concentration is uncertain.

VIII. Mutagenicity

Heptachlor is metabolized principally to its epoxide in mammalian systems (Davidow and Radomski 1953). Epoxides are often ultimate mutagens since they are to some degree reactive as electrophiles. Heptachlor and its epoxide have thus been investigated for mutagenicity in a variety of in vitro and in vivo tests.

Testing of heptachlor for mutagenic effects in most microbial systems has yielded negative results. Heptachlor was not mutagenic in the his^- reversion (Ames) assay for a variety of strains of *Salmonella typhimurium* in the absence (Marshall et al. 1976; Shirasu et al. 1976) or the presence of rat liver microsomes (Glatt et al. 1983; Moriya et al. 1983). It was likewise negative in reversion assays with auxotrophic strains of *Escherichia coli* (Moriya et al. 1983) and *Saccharomyces cerevisiae* (Gentile et al. 1982) as well as the *rec* assay with *Bacillus subtilis* (Shirasu et al. 1976). However one positive report has appeared indicating a 1.5- to 2-fold increase in reversions of two strains of *S. typhimurium* with technical but not commercial grade heptachlor following activation with rat liver microsomes (Gentile et al. 1982). Similar results were reported with technical grade heptachlor following activation with a plant extract system. Crebelli et al. (1986) failed to detect increased forward mutations, mitotic crossing-over, or chromosome malsegrations in the fungus *Aspergillus nidulans* treated with heptachlor epoxide. No breakage of the bacterial colicinogenic plasmid E1 was induced by incubation with heptachlor, while DNA alkylating agents were potent inducers of strand breaks (Griffin and Hill 1978).

Heptachlor was previously used in agricultural systems, which prompted testing for its ability to induce mutations in plants. These assays are based on the ability of mutagens to induce reverse mutations in the *waxy* locus of a strain of *Zea mays*. In situ treatment of *Z. mays* with 1.12 kg heptachlor ha^{-1} resulted in a twofold increase in reversions (Plewa and Wagner 1981). Heptachlor treatments (seeds soaked in solutions of up to 0.3% heptachlor) caused sticky chromosomes, metaphase-I multivalents, more abundant fragments and bridges, and a decreased mitotic index in somatic and meiotic chromosomes of several leguminous species (Jain and Sarbhoy 1987a,b), but the potential results of such changes were not fully addressed.

Heptachlor has been found to increase the incidence of hepatocellular carcinoma in the mouse (section IX); appropriately, hepatocytes were examined for heptachlor-induced mutagenesis. No unscheduled DNA synthesis (UDS) was induced in primary cultures of rat, mouse, or hamster hepatocytes at doses up to 10^{-4} M (Williams 1980; Probst et al. 1981). It also did not induce mutations in the hypoxanthine-guanine phosphoribosyl transferase locus in an adult rat liver epithelial cell line (Telang et al. 1981). However, UDS in SV-40 transformed human fibroblasts was slightly increased by 100 μM heptachlor and 10 μM heptachlor epoxide in the presence of rat liver microsomal enzymes (Ahmed et al. 1977).

Drosophila melanogaster injected with 5 μg ml^{-1} of heptachlor or heptachlor epoxide showed no increase in X-linked recessive lethals in postmeiotic germ-cell stages (Benes and Sram 1969).

In dominant lethal studies, male mice were treated orally or intraperitoneally with 7.5 or 15 mg kg^{-1} heptachlor/heptachlor epoxide (1:3), and then mated consecutively with virgin females over a 6-wk period. Two weeks after mating, females were sacrificed and numbers of implantation sites, resorption sites, and embryos recorded. The heptachlor:heptachlor epoxide mixture caused no change in preganancy rates, a very minor increase in dominant lethal indices, and no increase in early deaths in utero, which led the authors to conclude there was no indication of dominant lethality (Arnold et al. 1977).

Inhibition of mouse testicular DNA synthesis has been used as an *in vivo* assay for mutagenesis. Heptachlor, at 40 mg kg^{-1} orally, caused a slight, nonsignificant decrease in testicular DNA synthesis (Seiler 1977).

IX. Carcinogenicity

The widespread use of heptachlor as an insecticide and its persistence in both the environment and the tissues of man led to a series of studies designed to determine if heptachlor is a carcinogen. Positive results in a 2-yr rodent bioassay, but a lack of heptachlor-caused DNA damage, led to the suggestion that heptachlor and its epoxide are epigenetic carcinogens of the promoter type (Williams 1983). In vitro and in vivo studies have been conducted to test this hypothesis.

Chronic, high-dose feeding studies of rodents remain the standard assay for carcinogenicity of test compounds. Five major studies have been carried out with heptachlor and heptachlor epoxide, but few have appeared in the conventional literature. Reviews of these studies have been published by Epstein (1976) and the WHO (1984b), and the histologic findings have recently been reevaluated by Reuber (1987).

An FDA bioassay conducted in 1965 consisted of feeding C3H mice (100 each male and female) 0 or 10 μg g^{-1} of heptachlor or heptachlor epoxide for 2 yr (Epstein 1976; Reuber 1977b). There was high mortality in the heptachlor epoxide-treated mice. Although control groups had high spontaneous rates of hepatocellular carcinoma, both heptachlor and heptachlor epoxide caused about a twofold increase in this tumor for both male and female mice. There was no increase in any other tumor type.

Another mouse strain, the Charles River CD-1, was used in a study conducted by the International Research and Development Corporation (IRDC) for the Velsicol Chemical Corporation (Epstein 1976; Reuber 1987). Male and female mice were fed a mixture of 75% heptachlor epoxide and 25% heptachlor at 0, 1, 5, or 10 µg g^{-1} for 18 mon. Mortality was high in all groups and significant numbers of mice were lost to analysis due to decomposition. A dose-dependent effect was seen for the induction of hepatocellular carcinoma in both sexes with significant increases seen at 5 and 10 µg g^{-1} in males and at 10 µg g^{-1} in females. Necrosis of the liver was a common finding. Again, the only increases in tumors were those of the liver.

The National Cancer Institute (NCI 1977b) used the B6C3F1 mouse in a chronic heptachlor feeding study. The initial dose of heptachlor was changed during the experiment, due to toxicity, resulting in time-weighted average doses of 25.7 and 51.3 µg g^{-1}. Fifty mice of each sex were used in each dose group and compared to 10 concurrent controls. The only significant increase in tumors was for hepatocellular carcinoma in the high dose groups. Liver carcinoma increased in males from 26% in matched controls to 72% in the high dose group ($p < 0.001$ using the Fisher exact test). The tumors were classified from well-differentiated to very anaplastic. One pulmonary metastasis was noted in the high dose group.

Studies in rats have been much less conclusive. Two studies were conducted by the Kettering Laboratory in CFN rats using dietary doses of heptachlor from 0 to 10 µg g^{-1} (Epstein 1976; Reuber 1987). The overall occurrence of tumors in treated animals (all experimental groups considered together) was slightly increased in both studies with most of the tumors occurring in the endocrine organs. No preferential tumor sites were observed. One study showed an increase in liver tumors from 0% in controls to 8.4% in combined male and female treated groups while the other study found no induction of liver tumors.

Cabral et al. (1972) examined the effects of 10 mg of heptachlor kg^{-1} given in five administrations by gavage on alternate days to 95 suckling Wistar rats beginning at 10 d of age. Tumor incidences in treated and control male rats at 2 yr were similar. Overall tumor incidence in females increased from four in 27 control rats (15%) to nine in 28 treated rats (32%). The four control rats developed a total of four tumors (including two mammary tumors) and the nine treated rats developed 12 tumors, including five of the breast. No liver tumors were detected.

The NCI (1977b) heptachlor bioassay used 50 each male and female Osborne-Mendel rats in high and low dose groups. Technical grade heptachlor (consisting of 74% haptachlor and the rest essentially all α-chlordane) was fed to rats in time-weighted average doses (initial doses altered due to toxicity) of 38.9 and 77.9 µg g^{-1} for 111 weeks. Of the female rats, 55% and 74% of the high- and low-dose groups, respectively, survived the entire treatment period compared to 70% of the matched controls. The corresponding numbers for male rats were 58% and 62% compared to 70% for matched controls. Data from control rats in similar, concurrently run bioassays on other insecticides were pooled with data from the control rats of the heptachlor study resulting in 60 male and 60 female controls for additional statistical analysis. Body weight was reduced in the high dose males and, to a lesser extent, in high dose females, but was not affected in the low-dose groups. No hepatocellular carcinomas were diagnosed among the 187 heptachlor-treated rats examined. Neoplastic nodules of the liver were seen in both control and treated rats but no significant dose-dependent trend was found. There was a significant excess of lesions interpreted as follicular cell carcinomas of the thyroid in the treated male and female rats, although there was a significant decrease in C-cell thyroid tumors in treated animals. There were no other significant findings in rats.

Reuber (1987) recently reevaluated the histology of this study and came to different conclusions regarding hepatocellular carcinoma incidence. Although he found hyperplastic nodules in both control and treated groups, no hepatocellular carcinoma was diagnosed in the controls but was found in six of 42 (14%) and three of 37 (8%) in low and high dose groups, respectively, in male rats and two of 45 (4%) and five of 37 (14%) in female rats. It should be pointed out that there is a difference in terminology between the two reports. Neoplastic nodules were diagnosed in the NCI study, while Reuber refers to hyperplastic nodules. In the summary of Reuber's paper, he identifies neoplasms at many sites in the low and high dose rats; however, in his comments he states "The incidence of benign and malignant neoplasms was so high in control rats, particularly the females, that differences between the controls and heptachlor-treated rats often could not be detected."

Heptachlor is considered carcinogenic in the mouse liver despite deficiencies in study design in the experiments available for review. There is a possibility of carcinogenicity in the rat, but the differences of opinion between different observers regarding hepatocellular carcinomas leaves making a final decision difficult for a reviewer. The significance of all these findings is questionable, in light of the lack of a chronic feeding study to determine a maximum tolerated dose (MTD). The NCI (1977b) conducted an experiment on a small number of animals in order to estimate a MTD; however, this was obviously insufficient in view of the large number of animals lost early in the study before doses were adjusted downward. In addition, none of these studies were done using pure heptachlor. Indeed, the NCI bioassay, the most thoroughly conducted of the studies, used technical grade heptachlor containing approximately 25% α-chlordane. α-chlordane was identified as a carcinogen in the B6C3F1 mouse liver (NCI 1977a). Thus, it is important to determine the carcinogenic properties of heptachlor in the absence of significant contamination.

The assessment of human carcinogenic risk from compounds which cause tumors predominately in the livers of mice, particularly the B6C3F1 strain, has been the subject of much controversy due to the high spontaneous rate of tumors in these mice (Maronpot et al. 1987). The elegant work of Reynolds et al. (1987) provides insight into the mechanism of such tumor induction and the potential for distinguishing between promotion of background tumors versus a genotoxic pathway of tumor initiation.

DNA from spontaneous mouse liver tumors induced transformation of NIH/3T3 fibroblasts in culture (Fox and Watanabe 1985). Dominant transforming genes were detected in three of ten adenomas and 14 of 17 carcinomas from spontaneous liver tumors in B6C3F1 mice (Reynolds et al. 1987). The transforming gene was identified as activated Harvey-*ras* (H-*ras*) in all three adenomas and in 12 of 14 carcinomas (Reynolds et al. 1987). More importantly, all of the activated H-*ras* genes were shown to have a mutation in codon 61. Liver tumors induced by furan and furfural also contained a high percentage of transforming genes, mostly H-*ras*, but significantly some Kirsten-*ras*, an oncogene not detected in the spontaneous tumors. In addition, a broad spectrum of activating point mutations in H-*ras* was found in contrast to the specificity seen in spontaneous tumors. Thus, chemicals which induce H-*ras* or other oncogene activation independent of H-*ras* codon 61 activation in the mouse liver might be considered to have genotoxic properties. Such an examination could be conducted on heptachlor-induced mouse liver tumors. Indeed, this type of analysis is currently being done for chlordane, an insecticide related to heptachlor with similar carcinogenic properties (MW Anderson, NIEHS, personal communication). Additional information on specific H-*ras* mutations could be obtained from tumors previously generated in the NCI and other chronic feeding bioassays through the use of the polymerase chain reaction technique which amplifies specific gene sequences even on formalin-fixed tissues (Mullis and Faloona 1987). It must be noted that detection of mutations predominately in codon 61 of H-*ras* would not rule out a genotoxic mechanism of tumor induction; the powerful carcinogen *N*-nitrosodiethylamine (DEN) induces mutations primarily in codon 61 of the B6C3F1 mouse (Stowers et al. 1988). Evaluation of DNA adduct formation or cytotoxicity in this case might suggest possible carcinogenic mechanisms.

In analogy to skin carcinogenesis, Pitot et al. (1978) have shown that hepatocarcinogenesis can occur in two stages, initiation and promotion. Initiating agents are characterized by their ability to interact with DNA, causing irreversible changes. Promotion involves the selection and clonal expansion of initiated cells. Promotion can occur only after initiation, and is reversible. Identification of a promotion mechanism for heptachlor may have important regulatory implications since threshold levels for epigenetic tumor promoters are likely to exist (Pitot et al. 1987). Since heptachlor has been shown to be generally negative in mutagenesis tests, and caused liver cancer primarily in mice with a high spontaneous rate of such tumors, Williams and Numoto (1984) investigated the ability of heptachlor to act as a liver tumor promoter. B6C3F1 mice were initiated with a 2-wk exposure to DEN, a DNA alkylating agent, in the drinking water at 20 mg L^{-1}. This was followed by feeding with 5 or 10 μg technical grade heptachlor g^{-1} food over 25 wk. Initiated mice not fed heptachlor developed a 40% incidence of hepatocellular carcinoma, while those receiving heptachlor had an 80% incidence of these lesions. Mice exposed to DEN alone developed neoplasms of the stomach and lung, but heptachlor did not increase the incidence of these. Heptachlor administered for 25 wk without prior initiation did not result in an increase in neoplasms. In addition, heptachlor given for 25 wk before initiation with DEN did not result in enhanced production of neoplasms. Thus it was concluded that heptachlor acts as an epigenetic liver tumor promoter enhancing the yield of neoplasms from previously initiated cells. In B6C3F1 mice, which have a high spontaneous rate of hepatocellular carcinoma (indicating a high background of initiated cells), one would predict increased liver tumors following treatment with promoting agents. It should be noted that the study by Williams and Numoto (1984) did not result in increased tumor incidence with heptachlor alone, although the NCI (1977b) bioassay did. The difference would appear to be due to the duration of heptachlor administration, 25 wk in the former and 90 wk in the latter. Thus, prolonged exposure appears to be required and the effects of heptachlor exposure may be reversible after termination of short-term exposure. Testing the ability of heptachlor to act as a promoter in animals having a very low incidence of spontaneous hepatocellular carcinoma has not been reported but could provide evidence as to whether heptachlor acts as a complete carcinogen.

In further support of the hypothesis that heptachlor acts via an epigenetic mechanism, Telang et al. (1982) have shown that heptachlor produced effects similar to other tumor promoters by inhibiting intercellular communication between cultured liver epithelial cells. The skin tumor promoter 12-O-tetradecanoyl phorbol-13-acetate (TPA) (Yotti et al. 1979) and the liver tumor promoters 4,4'-dichlorodiphenyltrichloroethane (DDT) (Peraino et al. 1975; Williams and Numoto 1984) and phenobarbital (Ruch and Klaunig 1986) all appear to act by epigenetic mechanisms and have been found to inhibit intercellular communication in cell culture.

X. Teratogenicity

Heptachlor epoxide has been detected in the fat and blood of stillborn infants (Curley et al. 1969) as well as in umbilical cord blood of newborn infants (D'Ercole et al. 1976). This demonstrates transplacental transfer to the fetus and thus the potential for teratogenic effects. A variety of avian and mammalian studies have been conducted, and an accidentally exposed human population examined for evidence of teratogenesis.

Several groups have reported a 10–40% reduction in hatchability of chicken eggs injected with heptachlor or heptachlor epoxide; however, no abnormal chicks resulted (Dunachie and Fletcher 1969; McLaughlin et al. 1963; Smith et

Table 6. Maximum permissible concentrations of heptachlor

Source	Residue Concentration	
Inhalation		
Air[a]	0.5 mg m^{-3}	8-hr TWA
Contact		
Skin[a]	0.5 mg m^{-3}	8-hr TWA air concentration, allowing for cutaneous absorption
Ingestion		
Water[b]	0.1 µg L^{-1}	
Raw Agricultural Commodities[a]	0-0.1 mg kg^{-1} product	
Vegetable crops[c]	0.05 mg kg^{-1}	
Milk[c]	0.15 mg kg^{-1} (lw)	
Citrus fruit[c]	0.01 mg kg^{-1}	
"Food"[c]	0-0.5 µg kg^{-1} body wt	Acceptable Daily Intake (ADI)

[a] USPHS 1987.
[b] WHO 1984a.
[c] FAO/WHO 1971 (as cited in WHO 1984b).

al. 1970). Japanese and bobwhite quail fed diets containing 10 and 50 mg heptachlor kg^{-1} exhibited normal egg-laying and no adverse effects on chicks (Shellenberger et al. 1966).

Studies in mammals have been carried out for the Velsicol Chemical Corporation and are summarized in a World Health Organization report (WHO 1984b). Rats fed exclusively on diets containing heptachlor and/or heptachlor epoxide throughout three generations had a slight increase in pup mortality but no congenital malformations. In a two-generation study in beagle dogs, the highest dietary levels of heptachlor epoxide tested, 10 mg kg^{-1}, produced high pup mortality, but congenital malformations were absent at all doses. Oral treatment of pregnant rabbits with 5 mg heptachlor epoxide kg^{-1} d^{-1} during days 6-11 of gestation did not produce adverse effects on the fetuses. In cattle which ate heptachlor-contaminated feed, no difference in heptachlor epoxide residues were found in tissues of 31 spontaneously aborted cattle fetuses relative to nonaborted fetuses (Macklin and Ribelin 1971).

For 27-29 months during 1980-1982, the dairy milk supply of Oahu, Hawaii, was contaminated with heptachlor epoxide at levels of 1.2-5.0 µg g^{-1}, resulting in a four-fold increase in mean values of heptachlor epoxide in human milk (Smith 1982). An earlier study of concentrations in blood serum and breast milk indicated residue levels in the Oahu population similar to those in other areas of the US (Takahashi et al. 1981). Le Marchand et al. (1986) examined hospital records for incidences of 23 congenital malformations and found no differences in rates between the exposed population and rates in the same population before

Table 7. Heptachlor LD_{50}s in laboratory animals

Animal	Route	Dosage (mg kg^{-1})	Reference
chicken	oral	62	Sherman and Ross 1961
guinea pig	oral	116	Fitzhugh 1950
hamster	oral	160	Cabral et al. 1979
mouse	oral	68	Fitzhugh 1950
rabbit	dermal	2000[a]	Lehman 1952
	oral	80–90	Gleason et al. 1969
rat	dermal	195	Gaines 1969
	oral	22	Mampe 1987
		90	Lehman 1948
		100	Gaines 1969

[a] Single application, powder form.

exposure, or in rates in other unexposed populations (other Hawaiian islands and the total US) at the concurrent time.

XI. Toxicology and Human Exposure

Because of the numerous effects that heptachlor has on experimental animals, and the uncertainty about the effects of long-term exposure on humans, regulatory agencies have established maximum permissible levels of heptachlor exposure to humans in various settings. These levels are often expressed as a threshold limit value (TLV) or as a time-weighted average (TWA). TLV is defined as the allowable concentration of a substance to which most workers can be exposed without adverse effects. TWA is defined as the allowable concentration averaged over a normal 40-hr workweek (USPHS 1987). Some of the internationally recommended standards are outlined in Table 6.

A. Acute Toxicity

The acute toxicity of heptachlor has been examined in a number of species, and varies greatly. Some representative LD_{50} values are presented in Table 7. Because of the number of possible routes of heptachlor exposure, it is extremely important to examine toxicity via each exposure route. Most human exposures were believed to result from ingestion of foods of animal origin, i.e., dairy products, meat, fish, and poultry (Train 1976), but restricted agricultural use has made inhalation a potentially more important exposure route in homes treated with heptachlor as a termiticide. Heptachlor is also rapidly absorbed through intact skin, unlike DDT (Murphy 1986).

Heptachlor was less toxic to newborn rats than to adults (LD_{50} 531 mg kg^{-1} for neonates, compared to 71 mg kg^{-1} for adults) (Harbison 1973, 1975). Microsomal enzyme induction through PB administration increased acute toxicity to newborn rats (Harbison 1973). Likewise, the LD_{50} of heptachlor increased following turpentine pretreatment (112 mg kg^{-1} vs 70 mg kg^{-1} without turpentine pretreatment), whereas the heptachlor epoxide LD_{50} remained the same with or without turpentine pretreatment (Sperling and Ewenike 1972).

The best overall review of acute toxicity, by both oral and dermal routes, is the work of Gaines (1960, 1969). In the 1969 study, Gaines tested 19 chlorinated hydrocarbon pesticides in adult Sherman rats and found heptachlor to be the seventh most potent. It is interesting that heptachlor had a lower LD_{50} than chlordane, with which it is often contaminated. In addition to studying LD_{50} values, the author also assessed the LD_1, which many believe is a better measure of toxicity than the LD_{50}. The data indicated a linear dose-response relationship between the 1st and 50th percentiles. The author also developed a figure for what was termed "the lowest dose to kill a single rat," which is useful in determining comparative toxicity. Heptachlor was peculiar in being more toxic to males than females, which is unusual with most substances and certainly with most pesticides.

While it is true that acute toxicities and LD_{50} values have been well-studied in many animal species, and particularly in rats, the testing of toxicity in intermediate-duration studies (7 d through 6 mon) are mostly absent in the literature. An extensive review has failed to reveal any well-performed or controlled studies in the 3–6 mon time interval, where it would be possible to establish maximum tolerated doses. This is especially important when assessing carcinogenicity studies (cf. section IX). In an attempt to make some judgment of the intermediate-duration toxicity of heptachlor, one must turn to long-term studies designed mostly for carcinogenic effect (NCI 1977b; Reuber 1987).

Increased mortality in rats, mice, and dogs has been observed in intermediate stages of longer-term studies. A 6-wk study followed by 2 wk of observations with no dosing was reported as part of the NCI (1977b) study. The results reported were somewhat inadequate, but a no-effect level was suggested for rats at 8 mg kg^{-1} d^{-1}, and for mice at 6 mg kg^{-1} d^{-1}. In most of the long-term studies, which of course were designed primarily for tumor study, many rats died, and many of the studies cannot really be evaluated for intermediate-duration toxicity (cf. section IX).

There is a lack in the study of sub-acute to sub-chronic toxicity of heptachlor and heptachlor epoxide. A well-controlled 90 d study of heptachlor in rats and mice would probably be an advantage to assess carcinogenic potential more accurately.

The earliest manifestations of acute heptachlor poisoning arise from CNS stimulation, although the exact mechanism is not completely understood (cf. section 7). In fact, convulsions often precede any of the other less serious signs of poisoning (Murphy 1986). Other symptoms of acute heptachlor intoxication include profound hypothermia, arrythmias, and cardiovascular depression (Hrdina et al. 1974; Joy 1976).

B. Chronic Human Exposure

Because of their persistence and tendency to accumulate in fat depots, there has been a great deal of concern over the long-term, latent health effects of heptachlor, heptachlor epoxide and other organochlorine insecticides in man. Numerous studies have found detectable amounts of heptachlor epoxide in the majority of the general population studied. But as will be highlighted below, there exists no evidence to date that chronic low-level exposure to heptachlor is dangerous to human health.

Because of its propensity to concentrate in fats and its potential as a source of exposure to infants, heptachlor residue analysis has often been performed on breast milk. In fact, humans tend to concentrate pesticides to a much greater extent in their milk than do cows. Luquet et al. (1974) reported a ratio of 4–5:1 for heptachlor residues in human milk compared to cow milk, in samples from the same region. A 1965 study found traces of heptachlor epoxide, but no heptachlor, in 100% of the milk samples (Egan et al. 1965). Two years later, a Canadian study (Ritcey et al. 1972) again found only heptachlor epoxide, with undetectable amounts of the parent compound heptachlor.

Following the implementation of restricted use policies in 1972, a 1973 study (Jonsson et al. 1977) found heptachlor in 6% and heptachlor epoxide in 24% of samples obtained from the greater St. Louis, Missouri area. Concentrations of heptachlor in the positive samples were in the low ng g^{-1} range. A 1977 survey of Hawaiian women, exposed to heptachlor both as a termiticide and as a pesticide used on pineapple, revealed similar residue concentrations in breast milk (35 ng g^{-1} average residue) (Takahashi et al. 1981). Heptachlor residues were detected in 12% of Finnish human milk samples collected in 1984–1985, with a mean concentration of 0.02 and 0.05 mg kg^{-1} (lw) for heptachlor and heptachlor epoxide, respectively (Mussalo-Rauhamaa et al. 1988).

The most extensive study to date (Savage et al. 1981) utilized over 1,400 samples obtained throughout the US to determine both pesticide concentrations and the geographical distribution of exposure. Heptachlor was found in <2% of all samples, but traces of heptachlor epoxide were found in 63% of the samples. Geographically, samples from the southeastern US were the most commonly contaminated. The fact that proportionately more homes in this region are treated for termite control than in other US areas may explain this difference. Although the majority of positive samples contained only trace amounts, there were some instances of children being exposed to elevated doses of heptachlor by nursing. Mussalo-Rauhamaa et al. (1988) calculated that human infants should not be exposed to milk containing >0.1 mg heptachlor kg^{-1} milk fat.

The World Health Organization and the Food and Agriculture Organization have recommended 0.5 µg kg^{-1} as the acceptable daily intake (ADI) of heptachlor residues in foodstuffs (FAO/WHO 1979). It was calculated that the daily human intake of heptachlor in the US ranged from 0.29 to 0.64 µg d^{-1} during the period

1971–74 (Peirano 1980). This is lower than estimates for 1965 (2 µg d^{-1}) and 1970 (1 µg d^{-1}) (Duggan and Corneliussen 1972). Heptachlor epoxide can be found in small amounts in fish, poultry, meat, and dairy products, and in trace amounts in fruits, vegetables, oils and cereals. Dey and Parham (1988) recently examined over 1500 meat samples from eleven midwestern states and found that 2.2% of the samples contained violative levels of heptachlor residues. Given the restricted use of heptachlor at that time, these statistics were of interest, and the source of contamination was traced to tainted mash and heptachlor-treated seed grain used as fodder. Heptachlor has also been found in pharmaceuticals prepared from plant extracts. While tested foodstuff samples often contain detectable levels of heptachlor or heptachlor epoxide, the toxicological significance of such levels are generally not discussed. (See also section V.C, Bioaccumulation.)

C. Epidemiology

Isolated reports of people exposed only to heptachlor are, understandably, rare. The population-at-large is, however, commonly exposed to a multitude of marginally toxic substances whose effects, and certainly *combined* effects, are poorly understood. With these caveats, the most suitable groups for determining epidemiological effects of heptachlor on human health would be those involved in its manufacture and application.

In a large retrospective study conducted by Wang and MacMahon (1979a), 1,403 white males who were employed in the manufacture of heptachlor and chlordane between 1946 and 1976 were studied using individual follow-up questioning and Social Security records. The authors found no excess deaths from cancer, even among workers employed 20 or more years. This was confirmed by another group looking at workers from the same plant (Ditraglia et al. 1981).

In a study of another group of highly exposed individuals, pesticide applicators, no significant increases in the overall cancer rates were found (Saito et al. 1986, Wang and MacMahon 1979b). Although not significant, increased skin and bladder cancer rates were seen in the applicators. Ending in 1984, a further follow-up study of this same cohort revealed a standardized mortality rate for deaths due to lung cancer which was significantly less in chlordane and heptachlor applicators than in other pesticide applicators (MacMahon et al. 1988). No increases in any type of cancer was attributed to heptachlor. One must keep in mind that these applicators work with a variety of compounds.

After the 1980–1982 contamination of the milk supply on the Hawaiian island of Oahu caused by heptachlor-tainted cattle feed (Smith 1982), another group of people was available for scrutiny. Subsequent analysis of incidence rates for 23 congenital malformations failed to show any increase attributable to heptachlor exposure (Le Marchand 1986) (section X.C). Similarly, Stehr-Green et al. (1988) found no evidence of acute or sub-acute hepatic effects in individuals who had consumed heptachlor-contaminated milk.

While pesticide workers can be expected to be exposed to heptachlor on a continuous basis, the general population is also probably exposed continually, albeit in much smaller amounts. Two studies (Livingston and Jones 1981; Wright and Leidy 1982) showed detectable levels in homes that were improperly treated for termite control. Even in those homes where the termiticide was properly applied, detectable concentrations were found in the ambient air and on furniture and carpeting. Following subterranean application of heptachlor, Louis and Kisselbach (1987) found quantifiable concentrations of heptachlor in both living and non-living areas. Heptachlor concentrations in all the above studies were below the National Academy of Science's recommendation of 2.0 µg m^{-3} (National Research Council 1982) and the NIOSH standard of 0.5 mg m^{-3} (USPHS 1987).

XII. Recommendations

Current registration of heptachlor limits application to subterranean structural pest control, although the manufacturer (Velsicol Chemical Corporation) has voluntarily suspended sale of heptachlor-containing pesticides. This cessation has arisen out of concerns over the carcinogenic potential of residues in treated homes. Heptachlor is persistent in the environment; however, this restricted use does not appear to pose a widespread ecological concern if application methods contain the pesticide to the treated site. At or within sites of application, heptachlor may present a localized threat.

Studies using experimental animals suggest that multiple organ systems are adversely affected by high level heptachlor exposure. Significant differences between the ways that humans and other mammals metabolize heptachlor suggest that efforts should be focused on establishing the pharmacokinetic parameters in humans. Further work should be conducted to determine whether heptachlor or heptachlor epoxide are epigenetic tumor promoters and not genotoxic carcinogens. This should include establishment of MTDs, chronic feeding studies in animals other than mice and rats, initiation/promotion protocols in animals with a low rate of spontaneous liver neoplasms, as well as an analysis of activated oncogenes in heptachlor-induced mouse liver tumors. Chronic inhalation studies are also needed since this is the likely route of exposure for the human population in the 30 million residential structures treated with cyclodiene insecticides. Workers exposed during the manufacture of heptachlor and during its application should continue to be monitored for evidence of adverse health effects, including cancer.

In view of the controversy over the significance of mouse liver neoplasms as endpoints in evaluating the potential for harm to humans from chemicals, it is a formidable task to establish policies regulating human exposure. Definitive guidelines should await elucidation of the mechanisms which underlie the carcinogenic properties of this compound. In the absence of data to the contrary, however, it appears that continued use of heptachlor as a subterranean termiticide, with assurance that is can be applied in a correct and effective manner, poses limited human health hazard.

Summary

The chlorinated cyclodiene heptachlor was registered in 1952 as an agricultural and domestic insecticide. By early 1984, registration for all purposes, except subterranean termite control and for limited use in the control of fire ants, had been cancelled. This restriction of use arose primarily from concerns over the environmental persistance and bioaccumulation potential of the organochlorine pesticides. Currently, sale of heptachlor has been voluntarily suspended over questions about its carcinogenic potential, and the absence of safe and effective application methods.

As a persistent organochlorine pesticide, heptachlor residues are detected in all components of the environment. In historical use, heptachlor was directly applied to terrestrial systems, while air and water were secondarily contaminated via volatilization and land run-off, respectively. Within each environmental compartment, heptachlor undergoes a variety of metabolic and abiotic transformations. In vivo studies indicate that heptachlor epoxide is the predominant metabolite, formed as a product of the mixed-function oxidase system, while 1-hydroxychlordene is the major soil metabolite. For quantification, heptachlor and its metabolites are extracted from air, soil and sediment, water, or biological materials using various organic solvents and analyzed by gas chromatography or thin-layer chromatography.

Residue reports comprise most of the literature concerning the effects of heptachlor on the biota. In many such reports, toxic effects cannot be conclusively attributed to heptachlor exposure. Toxicity to organisms seems more dependent on acute exposure, while the chronic effects of low level exposure to heptachlor are poorly defined. Maximal terrestrial residues coincide with temporal and spatial proximity to application; peak residues in aquatic systems on the other hand, correlate to periods of maximum run-off. The lipophilic nature of both heptachlor and heptachlor epoxide results in the potential for significant bioaccumulation in all lipid-type compartments in the environment.

The toxic effects of heptachlor are not specific for any one organ system. The liver and the central nervous system are most significantly affected by heptachlor, although effects can also be seen in the reproductive, hematopoietic, immune, and renal systems. An important consideration is the relation of relevant environmental exposure levels to toxicity. The concentrations necessary to elicit results in laboratory experiments do not translate directly to the same results upon environmental exposure, nor do experimental laboratory animal models absolutely equate with native-state organisms or with humans.

Heptachlor and heptachlor epoxide generally lack mutagenic activity in microbial assays, mammalian cell culture systems, and in vivo studies. Chronic feeding studies showed increased incidence of hepatocellular carcinoma in mice, but were not conclusive in rats. In vivo and in vitro studies suggest that heptachlor functions as an epigenetic liver tumor promoter, and not a genotoxic carcinogen. There is no evidence for teratogenesis in either experimentally exposed animals or accidentally exposed humans. The evidence for carcinogenicity resulted in restrictions in the use of heptachlor imposed by the Environmental Protection Agency. Although data are limited, there is no evidence that long-term occupational or environmental exposure to heptachlor produces detrimental human health effects.

Acknowledgments

The Integrated Case Study approach to toxicological problems was developed by Dr. Curtis J. Richardson, professor in the School of Forestry and Environmental Studies at Duke University, under whose direction this review was conducted. Special thanks go to Dr. George D. Lumb for editorial assistance, and to Dr. M.B. Abou-Donia for initial guidance. Reviewers were supported by NIEHS training grant No. ES-07031.

References

Abalis IM, Eldefrawi ME, Eldefrawi AT (1985) High affinity stereospecific binding of cyclodiene insecticides and γ-hexachlorocyclohexane for γ-aminobutyric receptors of rat brain. Pestic Biochem Physiol 24:95–102.

Ahmed RE, Hart RW, Lewis NJ (1977) Pesticide induced DNA damage and its repair in cultured human cells. Mutat Res 42:161–174.

Albright LJ, Oloffs PC, Szeto SY (1980) Residues in cutthroat trout (*Salmo clarki*) and California newts (*Taricha torosa*) from a lake treated with technical chlordane. J Environ Sci Hlth Part B 15:333–350.

Alford-Stevens AL, Bellar TA, Eichelberger JW, Budde WL (1986) Accuracy and precision of determinations of chlorinated pesticides and polychlorinated biphenyls with automated interpretation of mass spectrometric data. Anal Chem 58:2022–2029.

Al-Omar MA, Al-Ogaily NH, Shebil DA (1986) Residues of organochlorine insecticides in fish from polluted water. Bull Environ Contam Toxicol 36:109–113.

Ambrus A, Hargitai E, Karoly G, Fulop A, Lantos J (1981a) General method for determination of pesticide residues in samples of plant origin, soil, and water. II. Thin layer chromatographic determination. J Assoc Off Anal Chem 64:743–748.

Ambrus A, Lantos J, Visi E, Csatlos I, Sarvari L (1981b) General method for determination of pesticide residues in samples of plant origin, soil, and water. I. Extraction and cleanup. J Assoc Off Anal Chem 64:733–742.

Amodio-Cocchieri R, Arnese A (1988) Organochlorine pesticide residues in fish from southern Italian rivers. Bull Environ Contam Toxicol 40:233–239.

Anderson DW, Raveling DG, Risebrough RW, Springer AM (1984) Dynamics of low-level organochlorines in adult cackling geese over the annual cycle. J Wildl Mgmt 48:1112-1127.

Anderson JM (1971) Sublethal effects and changes in ecosystems - assessment of the effects of pollutants on physiology and behavior. Proc R Soc London Ser B 177:307-320.

Arnold DW, Kennedy GL, Keplinger ML, Calandra JC (1977) Dominant lethal studies with technical chlordane, HCS-3260, and heptachlor:heptachlor expoxide. J Toxicol Environ Hlth 2:547-555.

Arruda J, Cringan M, Layher W, Kersh G, Bever C (1988) Pesticides in fish tissue and water from Tuttle Creek Lake Kansas. Bull Environ Contam Toxicol 41:617-624.

Arthur RD, Cain JD, Barrentine BF (1976) Atmospheric levels of pesticides in the Mississippi Delta. Bull Environ Contam Toxicol 15:129-134.

Atallah YH, Whitacre DM, Hoo BL (1979) Comparative volatility of liquid and granular formulations of chlordane and heptachlor from soil. Bull Environ Contam Toxicol 22:570-574.

Audus LJ (1960) Microbiological breakdown of herbicides in soil. In: Woodford EK, Sagar GR (eds) Herbicides and soil. Blackwell Scientific Publications, Ltd., Oxford. pp 1-19.

Bache CA, Gyrisco SN, Fertig SN, Huddleston EW, Lisk DJ, Fox FH, Trimberger GW, Holland RF (1960) Effects of feeding low levels of heptachlor epoxide to dairy cows on residues and off-flavors in milk. J Agric Food Chem 8:408-409.

Barbehenn KR, Reichel WL (1981) Organochlorine concentrations in bald eagles: brain/body lipid relations and hazard evaluation. J Toxicol Environ Hlth 8:325-330.

Barber RT, Warlen SM (1979) Organochlorine insecticide residues in deep sea fish from 2500m in the Atlantic Ocean. Environ Sci Technol 13:1146-1148.

Barquet A, Morgade C, Pfaffenberger CD (1981) Determination of organochlorine pesticides and metabolites in drinking water, human blood serum, and adipose tissue. J Toxicol Environ Hlth 7:469-479.

Beall ML Jr., Nash RG (1971) Organochlorine insecticide residues in soybean plant tops: root vs. vapor sorption. Agron J 63:460-464.

Beeman RW, Matsumura F (1981) Metabolism of *cis*- and *trans*-chlordane by a soil microorganism. J Agric Food Chem 29:84-89.

Belisle AA, Reichel WL, Locke LN, Lamont TG, Mulhern BM, Prouty RM, DeWolf RB, Cromartie E (1972) Residues of organochlorine pesticides, polychlorinated biphenyls, and mercury and autopsy data for bald eagles, 1969 and 1970. Pestic Monit J 6:133-138.

Benes V, Sram R (1969) Mutagenic activity of some pesticides in *Drosophila melanogaster*. Ind Med Surg 38:442-449.

Benson WW, Brock DW, Gabica J, Loomis M (1976) Pesticide and mercury levels in pelicans in Idaho. Bull Environ Contam Toxicol 15:543-546.

Bernstein G, Artz SA, Hasen J, Oppenheimer JH (1968) Hepatic accumulation of I-125-thyroxine in the rat: Augmentation by phenobarbital and chlordane. Endocrinology (Baltimore) 82:406

Bess HA, Hylin JW (1970) Persistence of termiticides in Hawaiian soils. J Econ Entomol 63:633-638.

Beyer WN, Gish CD (1980) Persistence in earthworms and potential hazards to birds of soil applied DDT, dieldrin and heptachlor. J Appl Ecol 17:295-307.

Bijleveld MFIJ, Goeldlin P, Mayol J (1979) Persistent pollutants in Audouin's Gull (*Larus audouinii*) in the western Mediterranean: A case study with wide implications? Environ Conserv 6:139-142.

Blackmore DK (1963) The toxicity of some chlorinated hydrocarbon insecticides to British wild foxes (*Vulpes vulpes*). J Comp Pathol Ther 73:391-409.

Blevins RD (1979) Organochlorine pesticides in gamebirds of eastern Tennessee. Water Air Soil Pollut 11:71-75.

Blus LJ, Lamont TG (1979) Organochlorine residues in six species of estuarine birds, South Carolina, 1971-75. Pestic Monit J 13:56-60.

Blus LJ, Neely BS Jr., Lamont TG, Mulhern BM (1977) Residues of organochlorines and heavy metals in tissues and eggs of brown pelicans, 1969-73. Pestic Monit J 11:40-53.

Blus LJ, Lamont TG, Neely BS Jr. (1979) Effects of organochlorine residues on eggshell thickness, reproduction, and populations status of brown pelicans (*Pelecanus occidentalis*) in South Carolina and Florida, 1969-76. Pestic Monit J 12:172-184.

Blus LJ, Pattee OH, Henny CJ, Prouty RM (1983) First records of chlordane-related mortality in wild birds. J Wildl Mgmt 47:196-198.

Blus LJ, Henny CJ, Lenhart DJ, Kaiser TE (1984) Effects of heptachlor- and lindane-treated seed on Canada geese. J Wildl Mgmt 48:1097-1111.

Blus LJ, Henny CJ, Krynitsky AJ (1985a) Organochlorine-induced mortality and residues in long-billed curlews from Oregon. Condor 87:563-565.

Blus LJ, Henny CJ, Krynitsky AJ (1985b) The effects of heptachlor and lindane on birds, Columbia Basin, Oregon and Washington, 1976-1981. Sci Total Environ 46:73-81.

Boddicker ML, Hugghins EJ, Richardson AH (1971) Parasites and pesticide residues of mountain goats in South Dakota. J Wildl Mgmt 35:94-103.

Bonderman DP, Slach E (1972) Appearance of 1-hydroxychlordene in soil, crops, and fish. J Agric Food Chem 20:328-331.

Bradshaw JS, Loveridge EL, Rippee KP, Peterson JL, White DA, Barton JR, Fuhriman DK (1972) Seasonal variations in residues of chlorinated hydrocarbon pesticides in the water of the Utah lake drainage system—1970 and 1971. Pestic Monit J 6:166-178.

Braun HE, Frank R (1980) Organochlorine and organophosphorus insecticides: Their use in eleven agricultural watersheds and their loss to stream waters in southern Ontario, Canada. Sci Total Environ 15:169-192.

Bresch H, Arendt U (1977) Influence of different organochlorine pesticides on the development of the sea urchin embryo. Environ Res 13:121-128.

Bridges WR (1965) Effects of time and temperature on the toxicity of heptachlor and kepone to redear sunfish. In: Tarzwell CM (ed) Biological problems in water pollution, 3rd Seminar 1962. US Dept HEW, Public Health Services, Atlanta, Georgia. pp 247-249.

Brodie JE, Hichs WS, Richards GN, Thomas FG (1984) Residues related to agricultural chemicals in the groundwaters of the Burdekin river delta, North Queensland. Environ Pollut Ser B 8:187-215.

Brooks GT (1969) The metabolism of diene-organochlorine (cyclodiene) insecticides. Residues Reviews 27:81-138.

Bulkley RV, Leung SYT, Richard JJ (1981) Organochlorine insecticide concentrations in fish of the Des Moines River, Iowa, 1977–78. Pestic Monit J 15:86–89.

Burnham AK, Calder GV, Fritz JS, Junk GA, Svec HV, Willis R (1972) Identification and estimation of neutral organic contaminations. Anal Chem 44:139–142.

Burns BG, Peah ME, Stiles DA (1975) Organochlorine pesticide residues in a farming area, Nova Scotia. Pestic Monit J 9:34–38.

Burrage RH, Saha JG (1967) Insecticide residues in spring wheat plants grown in the field from seed treated with aldrin or heptachlor. Can J Plant Sci 47:114–115.

Burrage RH, Saha JG (1972) Insecticide residues in pheasants after being fed on wheat seed treated with heptachlor and ^{14}C-lindane. J Econ Entomol 65:1013–1017.

Butler PA (1963) Commercial fisheries investigations. US Fish Wildl Serv Circ 167:11–25.

Cabral JR, Testa MC, Terracini B (1972) Lack of long-term effects of the administration of heptachlor to suckling rats. Tumori 58:49–53.

Cabral JRP, Raitano R, Mollner T, Bronczyk S, Shubik P (1979) Acute toxicity of pesticides in hamsters. Toxicol Appl Pharmacol 48:A192.

Cain BW (1981) Nationwide residues of organochlorine compounds in wings of adult mallards and black ducks, 1979–80. Pestic Monit J 15:128–134.

Campbell MA, Gyorkos J, Leece B, Homonko K, Safe S (1983) The effects of 22 organochlorine pesticides as inducers of the hepatic drug-metabolizing enzymes. Gen Pharmacol 14:445–454.

Carey AE, Gowen JA (1979) Pesticide application and cropping data from 37 states, 1972 – National Soils Monitoring Program. Pestic Monit J 12:198–208.

Carey AE, Wiersma GB, Tai H, Mitchell WG (1972) Organochlorine pesticide residues in soils and crops of the Corn Belt Region, United States – 1970. Pestic Monit J 6:369–376.

Carey AE, Wiersma GB, Tai H (1976) Pesticide residues in urban soils from 14 United States cities, 1970. Pestic Monit J 10:54–60.

Carey AE, Gowen JA, Tai H, Mitchell WG, Wiersma GB (1978a) Pesticide residue levels in soils and crops, 1971 – National Soils Monitoring Program (III). Pestic Monit J 12:117–136.

Carey AE, Gowen JA, Wiersma GB (1978b) Pesticide application and cropping data from 37 states, 1971 – National Soils Monitoring Program. Pestic Monit J 12:137–148.

Carey AE, Gowen JA, Tai H, Mitchell WG, Wiersma GB (1979a) Pesticide residue levels in soils and crops from 37 states, 1972 – National Soils Monitoring Program (IV). Pestic Monit J 12:209–229.

Carey AE, Douglas P, Tai H, Mitchell WG, Wiersma GB (1979b) Pesticide residue concentrations in soils of five United States cities, 1971 – Urban Soils Monitoring Program. Pestic Monit J 13:17–22.

Caro JH (1971) Accumulation of dieldrin and heptachlor on corn leaves in and around a treated field. J Agric Food Chem 19:78–80.

Carter FL, Stringer CA (1970) Residues and degradation products of technical heptachlor in various soil types. J Econ Entomol 63:625–628.

Carter FL, Stringer CA, Heinzelman D (1971) 1-hydroxy-2,3-epoxychlordene in Oregon soil previously treated with technical heptachlor. Bull Environ Contam Toxicol 6:249–254.

Caster WO, Wade AE, Greene FE, Meadows JS (1970) Effect of different levels of corn

oil in the diet upon the rate of hexobarbital, heptachlor and aniline metabolism in the liver of the male white rat. Life Sci 9:181-190.
Castro TF, Yoshida T (1971) Degradation of organochlorine insecticides in flooded soils in the Philippines. J Agric Food Chem 19:1168-1170.
Chapman RA, Cole CM (1982) Observations on the influence of water and soil pH on the persistence of insecticides. J Environ Sci Hlth Part B 17:487-504.
Chawla RP, Kalra RL, Joia BS (1981) Absorption of residues of soil applied aldrin and heptachlor in potatoes. Indian J Entomol 43:266-271.
Clark DR Jr., Krynitsky AJ (1980) Organochlorine residues in eggs of loggerhead and green sea turtles nesting at Merritt Island, Florida – July and August 1976. Pestic Monit J 14:7-10.
Clark DR Jr., McLane MAR (1974) Chlorinated hydrocarbon and mercury residues in woodcock in the United States, 1970-71. Pestic Monit J 8:15-22.
Clark DR Jr., La Val RK, Krynitsky AJ (1980) Dieldrin and heptachlor residues in dead gray bats, Franklin County, Missouri – 1976 versus 1977. Pestic Monit J 13:137-140.
Clark DR Jr., La Val RK, Tuttle MD (1981) Estimating pesticide burdens of bats from guano analyses. Bull Environ Contam Toxicol 29:214-220.
Clark DR Jr., Bunck CM, Cromartie E, La Val RK (1983) Year and age effects on residues of dieldrin and heptachlor in dead gray bats, Franklin County, Missouri – 1976, 1977, and 1978. Environ Toxicol Chem 2:387-393.
Clawson SG, Baker MF (1959) Immediate effects of dieldrin and heptachlor on bobwhites. J Wildl Mgmt 23:215-219.
Cochrane WP, Greenhalgh R (1976) Chemical composition of technical chlordane. J Am Chem Soc 59:696-702.
Collins JA, Langlois BE (1968) Effect of DDT, dieldrin, and heptachlor on the growth of selected bacteria. Appl Microbiol 16:799-800.
Conney AH, Welch RM, Kuntzman R, Burns JJ (1967) Effects of pesticides on drug and steroid metabolism. Clin Pharmacol Ther (St. Louis) 8:2-10.
Crebelli D, Bellincampi D, Conti G, Conti L, Morpurgo G, Carere A (1986) A comparative study on selected chemical carcinogens for chromosomal malsegregation, mitotic crossing-over and forward mutation induction in *Aspergillus nidulans*. Mutat Res 172:139-149.
Crockett AB, Wiersma GB, Tai H, Mitchell WG, Sand PF, Carey AE (1974) Pesticide residue levels in soils and crops, FY-70 – National Soils Monitoring Program (II). Pestic Monit J 8:69-97.
Cromartie E, Reichel WL, Locke LN, Belisle AA, Kaiser TE, Lamont TG, Mulhern BM, Prouty RM, Swineford DM (1975) Residues of organochlorine pesticides and polychlorinated biphenyls and autopsy data for bald eagles, 1971-72. Pestic Monit J 9:11-14.
Curley A, Copeland M, Kimbrough RD (1969) Chlorinated hydrocarbon insecticides in organs of stillborn and blood of newborn babies. Arch Environ Hlth 19:628-632.
Cutkomp LK, Yap HH, Cheng EY, Koch RB (1971) ATPase activity in fish tissue homogenates and inhibitory effects of DDT and related compounds. Chem Biol Interact 3:439-447.
Davidow B, Radomski JL (1953) Isolation of an epoxide metabolite from fat tissues of dogs fed heptachlor. J Pharmacol Exp Ther 107:259-265.

Davis PW, Friedhoff JM, Wedeneyer GA (1972) Organochlorine insecticide, herbicide, and polychlorinated biphenyl (PCB) inhibition of NaK-ATPase in rainbow trout. Bull Environ Contam Toxicol 8:69–72.

Delany MF, Bell JU, Sundlof SF (1988) Concentrations of contaminants in muscle of the American alligator in Florida. J Wildl Dis 24:62–66.

Den Tonkelaar EM, Van Esch GJ (1974) No-effect levels of organochlorine pesticides based on induction of microsomal enzymes in short-term toxicity experiments. Toxicology 2:371–380.

D'Ercole AJ, Arthur RD, Cain JD, Barrentine BF (1976) Insecticide exposure of mothers and newborns in a rural agricultural area. Pediatrics 57:869–874.

De Smet KD (1987) Organochlorines, predators and reproductive success of the red-necked grebe in southern Manitoba. Condor 89:460–467.

DeWeese LR, McEwen LC, Hensler GL, Petersen BE (1986) Organochlorine contaminants in Passeriformes and other avian prey of the peregrine falcon in the western United States. Environ Toxicol Chem 5:675–693.

DeWitt JB, Menzie CM, Adomaitis VA, Reichel WL (1960) Pesticidal residues in animal tissues. Trans North Am Wildl Nat Resour Conf 25:277–285.

Dey BP, Parham GL (1988) An episode for heptachlor contamination in food animals. Arch Environ Hlth 43(2):202.

Ditraglia E, Brown DP, Namekata T, Iverson N (1981) Mortality study of workers employed at organochlorine pesticide manufacturing plants. Scand J Work Environ Hlth 7:(suppl. 4)140–146.

DouAbul AAZ, Al-Omar M, Al-Obaidy S, Al-Ogaily N (1987) Organochlorine pesticide residues in fish from the Shatt al-Arab river, Iraq. Bull Environ Contam Toxicol 38:674–680.

DouAbul AAZ, Al-Saad H, Al-Timari A, Al-Rekabi H (1988) Tigris-Euphrates delta: a major source of pesticides to the Shatt al-Arab river (Iraq). Arch Environ Contam Toxicol 17:405–418.

Dowd PF, Mayfield GU, Coulon DP, Graves JB, Newsom JD (1985) Organochlorine residues in animals from three Louisiana watersheds in 1978 and 1979. Bull Environ Contam Toxicol 34:832–841.

Duggan RE, Corneliussen PE (1972) Dietary intake of pesticide chemicals in the United States (III), June 1968–April 1970. Pestic Monit J 5:331–341.

Dunachie JF, Fletcher WW (1969) An investigation of the toxicity of insecticides to birds' eggs using the egg injection technique. Ann Appl Biol 64:409–423.

Dvorak M, Halacka K (1975) Ultrastructure of liver cells in pig at normal conditions and after administration of small doses of heptachlorine. Folia Morphol (Prague) 23:71–76.

Edwards CA (1966) Insecticide residues in soils. Residue Reviews 13:83–132.

Edwards CA, Thompson AR (1973) Pesticides and the soil fauna. Residue Reviews 45:1–79.

Edwards WR, Duzan RE, Siemers RJ (1983) Organochlorine insecticide residues and PCBs in tissues of woodcock, mourning doves, and robins from east-central Illinois, 1978–79. Bull Environ Contam Toxicol 31:407–414.

Egan H, Goulding R, Roburn J, Tatton J (1965) Organochlorine pesticide residues in human fat and human milk. Br Med J 2:66–69.

Eichelberger JW, Lichtenberg JJ (1971) Persistence of pesticides in river water. Environ Sci Technol 5:541–544.

Eisler R (1969) Acute toxicities of insecticides to marine decapod crustaceans. Crustaceana (Leiden) 16:307-310.

Eisler R (1970) Acute toxicities of organochlorine and organophosphorous insecticides to estuarine fishes. Tech Pap Bur Sport Fish Wildl US No. 46.

Ellenton JA, Brownlee LJ, and Hollebone BR (1985) Aryl hydrocarbon hydroxylase levels in herring gull embryos from different locations on the Great Lakes. Environ Toxicol Chem 4:615-622.

Elliot JE, Norstrom RJ, Keith JA (1988) Organochlorines and eggshell thinning in northern gannets (*Sula bassanus*) from eastern Canada, 1968-1984. Environ Pollut 52:81-102.

EPA (1987) Environmental News press release, August 11, 1987. Office of Public Affairs (A-107), US EPA, Washington, D.C.

Epstein SS (1976) Carcinogenicity of heptachlor and chlordane. Sci Total Environ 6:103-154.

Erney DR (1983) Rapid screening procedure for pesticides and polychlorinated biphenyls in fish: Collaborative study. J Assoc Off Anal Chem 66:969-973.

FAO/WHO (1971) Heptachlor. In: 1970 Evaluation of some pesticide residues in food. Food and Agriculture Organization of the United Nations, Rome. (as cited in WHO 1984b)

FAO/WHO (1979) Report of the joint meeting of the FAO panel of experts on pesticide residues and environment and the WHO expert group on pesticide residues held in Rome, Nov. 27-Dec. 6, 1978. FAO plant production and protection paper 15, pesticide residues in food – 1978. (as cited in Mussalo-Rauhamaa *et al.* 1988)

Ferguson DE (1964) Some ecological effects of heptachlor on birds. J Wildl Mgmt 28:158-163.

Feroz M, Khan MAQ (1979) Metabolism of ^{14}C heptachlor in goldfish (*Carassius auratus*). Arch Environ Contam Toxicol 8:519-531.

Fick GW (1977) Methods for evaluating insecticide effects on alfalfa growth. J Environ Qual 6:443-445.

Fitzhugh OG (1950) Reported by Cox AJ at US Food and Drug Administration Hearings of Week ending 9/12/1950. (as cited in Negherbon 1959)

Fleming WJ, Cain BW (1985) Areas of localized organochlorine contamination in Arizona and New Mexico. Southwest Nat 30:269-277.

Flickinger EL, Krynitsky AJ (1987) Organochlorine residues in ducks on playa lakes of the Texas Panhandle and eastern New Mexico. J Wildl Dis 23:165-168.

Foley RE, Batcheller GR (1988) Organochlorine contaminants in common goldeneye wintering on the Niagara River. J Wildl Mgmt 52:441-445.

Fox CJS, Chisolm D, Stewart DKR (1964) Effect of consecutive treatments of aldrin and heptachlor on residues in rutabaga and carrots and on certain soil arthropods and yield. Can J Plant Sci 44:149-156.

Fox GA, Kennedy SW, Norstrom RJ, Wigfield DC (1988) Porphyria in herring gulls: Biochemical response to chemical contamination of Great Lakes food chains. Environ Toxicol Chem 7:831-839.

Fox TR, Watanabe PG (1985) Detection of cellular oncogene in spontaneous liver tumors of B6C3F1 mice. Science (Washington, DC) 228:596-597.

Frank R (1981) Pesticides and PCB in the Grand and Saugeen river basins. J Great Lakes Res 7:440-454.

Frank R, Braun HE, Ishida K, Suda P (1976) Persistent organic and inorganic pesticide residues in orchard soils and vineyards of southern Ontario. Can J Soil Sci 56: 463–484.

Frank R, Van Hove Holdrinet M, Desjardine R, Dodge DP (1978) Organochlorine and mercury residues in fish from Lake Simcoe, Ontario 1970–1976. Environ Biol Fishes 3:275–286.

Frank R, Thomas RL, Holdrinet H, McMillan RK, Braun HE, Dawson R (1981) Organochlorine residues in suspended soils collected from the mouths of Canadian streams flowing into the Great Lakes 1974–1977. J Great Lakes Res 7:363–381.

Frank R, Braun HE, Van Hove Holdrinet M, Sirons GJ, Ripley BD (1982) Agricultural and water quality in the Canadian Great Lakes Basin: V. Pesticide use in 11 agricultural watersheds and presence in stream water 1975–1977. J Environ Qual 11:497–505.

Freeman HP, Taylor AW, Edwards WM (1975) Heptachlor and dieldrin disappearance from a field soil measured by annual residue determinations. J Agric Food Chem 23:1101–1105.

Fyfe RW, Campbell J, Hayson B, Hodson K (1969) Regional population declines and organochlorine insecticides in Canadian prairie falcons. Can Field-Nat 83:191–200.

Fyfe RW, Risebrough RW, Walker W II (1976) Pollutant effects on the reproduction of the prairie falcons and merlins of the Canadian prairies. Can Field-Nat 90:346–355.

Gabica J, Watson M, Benson WW (1974) Rapid gas chromatographic method for screening of pesticides. J Assoc Off Anal Chem 57:173–175.

Gaines TB (1960) The acute toxicity of pesticides to rats. Toxicol Appl Pharmacol 2:88–99.

Gaines TB (1969) Acute toxicity of pesticides. Toxicol Appl Pharmacol 14:515–534.

Gannon N, Bigger JH (1958) The conversion of aldrin and heptachlor to their epoxides in soil. J Econ Entomol 51:1–2.

Gannon N, Decker GC (1958) The conversion of heptachlor to its epoxide on plants. J Econ Entomol 51:3–7.

Gant DB, Eldefrawi ME, Eldefrawi AT (1987) Cyclodiene insecticides inhibit $GABA_A$ receptor-regulated chloride transport. Toxicol Appl Pharmacol 88:313–321.

Gasper GM, Kawatski JA (1972) Inhibition of dehydrogenase activity in mouse liver homogenates. Comp Biochem Physiol B: Comp Biochem 41:655–660.

Gawaad AAA, El-Gayar FH, Soliman AS, Rehab FL, Ali NM (1972) Effect of some soil insecticides on plants. I. On cotton, clover, bean and corn. Phytopathol Z 73:189–200.

Gentile JM, Gentile GJ, Bultmanm J, Sechriest R, Wagner ED, Plewa MJ (1982) An evaluation of the genotoxic properties of insecticides following plant and animal activation. Mutat Res 101:19–29.

Gesser HD, Chow A, Davies FC, Uthe JF, Reinke J (1971) The extraction and recovery of polychlorinated biphenyls (PCB) using porous polyurethane foam. Anal Lett 4:883–886.

Geyer H, Scheunert I, Korte F (1986) Bioconcentration potential of organic environmental chemicals in humans. Regul Toxicol Pharmacol 6:313–347.

Ghiasuddin SM, Matsumura F (1982) Inhibition of gamma-aminobutyric acid-induced chloride uptake by gamma-BHC and heptachlor epoxide. Comp Biochem Physiol C: Comp Pharmacol 73:141–144.

Gilman AP, Fox GA, Peakall DB, Teeple SM, Carroll TR, Haymes GT (1977) Reproductive parameters and egg contaminant levels of Great Lakes herring gulls. J Wildl Mgmt 41:458–468.

Gish CD (1970) Organochlorine insecticide residues in soils and soil invertebrates from agricultural lands. Pestic Monit J 3:241–252.

Glatt H, Jung R, Oesch F (1983) Bacterial mutagenicity investigation of epoxides: drugs, drug metabolites, steroids, and pesticides. Mutat Res 11:99–118.

Gleason MN, Gosselin RE, Hodge HC, Smith RP (1969) Clinical Toxicology of Commercial Products, 3rd Ed. Williams and Wilkins, Baltimore, MD. Section III, p 120.

Glotfelty DE, Taylor AW, Turner BC, Zoller WH (1984) Volatilization of surface-applied pesticides from fallow soil. J Agric Food Chem 32:638–643.

Gosselin RE, Hodge HC, Smith RP, Gleason MN (1976) Clinical Toxicology of Commercial Products, 4th Ed. Williams and Wilkins, Baltimore, MD. Section III, pp 168–169.

Gowen JA, Wiersma GB, Tai H, Mitchell WG (1976) Pesticide levels in hay and soils from nine states, 1971. Pestic Monit J 10:114–116.

Green DR, Stull JK, Heesen TC (1986) Determination of chlorinated hydrocarbons in coastal waters using a moored *in situ* sampler and transplanted live mussels. Mar Pollut Bull 17:324–329.

Greenberg RE, Heye PL (1971) Insecticide residues in little blue herons. Wilson Bull 83:95–97.

Greene FE (1972) Interactions of cyclodiene insecticides with components of the hepatic mixed function oxidase system from male and female rats. Toxicol Appl Pharmacol 22:309–310.

Greichus YA, Greichus A, Reider EG (1968) Insecticide residues in grouse and pheasant of South Dakota. Pestic Monit J 2:90–92.

Greichus YA, Greichus A, Emerick RJ (1973) Insecticides, polychlorinated biphenyls and mercury in wild cormonants, pelicans, their eggs, food and environment. Bull Environ Contam Toxicol 9:321–328.

Greichus YA, Gueck BD, Ammann BD (1978) Organochlorine insecticide, polychlorinated biphenyl, and metal residues in some South Dakota birds, 1975–76. Pestic Monit J 12:4–7.

Griffin DE III, Hill WE (1978) *In vitro* breakage of plasmid DNA by mutagens and pesticides. Mutat Res 52:161–169.

Grob K, Zurcher F (1976) Stripping of trace organic substances from water: Equipment and procedure. J Chromatogr 117:285–294.

Guengerich FP (1988) Roles of cytochrome P-450 enzymes in chemical carcinogenesis and cancer chemotherapy. Cancer Res 48:2946–2954.

Haake J, Kelley M, Keys B, Safe S (1987) The effect of organochlorine pesticides as inducers of testosterone and benzo-a-pyrene hydroxylases. Gen Pharmacol 18:165–169.

Hall RJ (1980) Effects of environmental contaminants on reptiles: A review. Spec Sci Rep: Wildl US Fish Wildl Serv No. 228.

Hall RJ, Kaiser TE, Robertson WB Jr., Patty PC (1979) Organochlorine residues in eggs of the endangered American crocodile (*Crocodylus acutus*). Bull Environ Contam Toxicol 23:87–90.

Harbison RD (1973) DDT, heptachlor, chlordane and parathion toxicity in adult, newborn and phenobarbital treated new-born rat. Toxicol Appl Pharmacol 25:472–473.

Harbison RD (1975) Comparative toxicity of some selected pesticides in neonatal and adult rats. Toxicol Appl Pharmacol 32:443–446.

Harris CR (1971) Influence of temperature on the biological activity of insecticides in soil. J Econ Entomol 64:1044–1049.

Harris CR (1972) Factors influencing the biological activity of technical chlordane and some related components in soil. J Econ Entomol 65:341-347.

Harris CR, Lichtenstein EP (1961) Factors affecting the volatilization of insecticidal residues from soils. J Econ Entomol 54:1038-1045.

Harris CR, Sans WW (1971) Insecticide residues in soils on 16 farms in southwestern Ontario—1964, 1966, and 1969. Pestic Monit J 5:259-267.

Harris CR, Sans WW (1972) Behavior of heptachlor epoxide in soil. J Econ Entomol 65:336-341.

Haseltine SD, Mulhern BM, Stafford C (1980) Organochlorine and heavy metal residues in black duck eggs from the Atlantic flyway, 1978. Pestic Monit J 14:53-57.

Haseltine SD, Heinz GH, Reichel WL, Moore JF (1981) Organochlorine and metal residues in eggs of waterfowl nesting on islands in Lake Michigan off Door County, Wisconsin, 1977-78. Pestic Monit J 15:90-97.

Hashemy-Tonkabony SE, Langaroodi FA (1976) Detection and determination of chlorinated pesticide residues in Caspian sea fish by gas-liquid chromatography. Environ Res 12:275-280.

Havera SP, Duzan RE (1986) Organochlorine and PCB residues in tissues of raptors from Illinois, 1966-1981. Bull Environ Contam Toxicol 36:23-32.

Heath RG (1969) Nationwide residues of organochlorine pesticides in wings of mallards and black ducks. Pestic Monit J 3:115-123.

Heath RG, Hill SA (1974) Nationwide organochlorine and mercury residues in wings of adult mallards and black ducks during the 1969-70 hunting season. Pestic Monit J 7:153-164.

Henderson C, Johnson WL, Inglis A (1969) Organochlorine insecticide residues in fish. Pest Monit J 3:145-171.

Henny CJ, Bean JR, Fyfe RW (1976) Elevated heptachlor epoxide and DDE residues in a merlin that died after migrating. Can Field-Nat 90:361-363.

Henny CJ, Blus LJ, Stafford CJ (1983) Effects of heptachlor on American kestrels in the Columbia Basin, Oregon. J Wildl Mgmt 47:1080-1087.

Hermanson HP, Gunther FA, Anderson LD, Garber MJ (1971) Installment application effects upon insecticide residue content of a California soil. J Agric Food Chem 19:722-726.

Hernandez LM, Rico MC, Gonzales MJ, Montero MC, Fernandez MA (1987) Residues of organochlorine chemicals and concentrations of heavy metals in Ciconiiform eggs in relation to diet and habitat. J Environ Sci Hlth Part B 22:245-258.

Herzel F (1972) Organochlorine insecticides in surface waters in Germany. 1970 and 1971. Pestic Monit J 6:179-187.

Heys AE, Duncan WP, Perry WC, Ebert DA, Radolovich G, Haile CL (1979) Synthesis of carbon-14 and carbon-13 labeled chlorinated polycyclic pesticides. J Labelled Compd & Radiopharm 16:295-306.

Hill EF, Heath RG, Spann JW, Williams JD (1975) Lethal dietary toxicities of environmental pollutants to birds. Spec Sci Rep: Wildl US Fish Wildl Serv No. 191.

Hiltibran RC (1974) Oxygen and phosphate metabolism of bluegill liver mitochondria in the presence of some insecticides. Trans Illinois State Acad Sci 67:228-237.

Holcome DW, Smith GS, Khan MF, Hallford DM, Rozman K (1988) Elimination of ^{14}C-heptachlor from body stores of lactating ewes treated with ovine growth hormone. J An Sci 66:2200-2208.

Holden AV (1973) Effects of pesticides on fish. In: Edwards CA (ed) Environmental pollution by pesticides. Plenum Press, London. pp 213–253.

Hrdina PD, Singhal RL, Peters DAV (1974) Changes in brain biogenic amines and body temperature after cyclodiene insecticides. Toxicol Appl Pharmacol 29:119.

Huber JT, Bishop JL (1961) Secretion of heptachlor epoxide in the milk of cows fed field-cured hay from soils treated with heptachlor. J Dairy Sci 45:79–81.

Hyman J (1949) Halogenated cyclopentadiene-hexachlorocyclopentadiene adducts. Chem Abstr 43:5796–5797.

Hyman J (1951) Polyhalogenated polycyclic hydrocarbons and insecticides. Chem Abstr 45:647.

IARC (1974) Some organochlorine pesticides. In: Monographs on the evaluation of carcinogenic risk of chemicals to man: Vol. 5. International Agency for Research on Cancer, Lyon, France. pp 173–191.

IRPTC (1982) Scientific reviews of Soviet literature on toxicity and hazards of chemicals, heptachlor. Centre of International Projects (GKNT, No. 3), Moscow. (as cited in WHO 1984b).

Ivie GW, Knox JR, Khalifa S, Yamamoto I, Casida JE (1972) Novel photoproducts of heptachlor epoxide, *trans*-chlordane and *trans*-nonachlor. Bull Environ Contam Toxicol 7:376–382.

Jain AK, Sarbhoy RK (1987a) Cytogenetical studies on the effect of some chlorinated pesticides I. Effect on somatic chromosomes of *Lens* and *Pisum*. Cytologia 52:47–53.

Jain AK, Sarbhoy RK (1987b) Cytogenetical studies on the effect of some chlorinated pesticides II. Effect on meiotic chromosomes of *Lens* and *Pisum*. Cytologia 52:55–61.

Javadi I, Hajari R (1986) A rapid TLC technique for determination levels of heptachlor in water. Toxicol Lett 31(suppl):191.

Jeffries DJ, Prestt I (1966) Post-mortems of peregrines and lanners with particular reference to organochlorine residues. Br Birds 59:49–64.

Johnson WW, Finley MT (1980) Handbook of acute toxicity of chemicals to fish and aquatic invertebrates. US Fish Wild Serv Resour Publ No. 137.

Jonsson V, Liu G, Armbruster J, Kettelhut LL, Drucker B (1977) Chlorohydrocarbon pesticide residues in human milk in greater St. Louis, Missouri 1977. Am J Clin Nutr 30:1106–1109.

Joy RM (1976) Convulsive properties of chlorinated hydrocarbon insecticides in the cat central nervous system. Toxicol Appl Pharmacol 35:95–106.

Kacew S, Singhal RL (1974) Effect of certain halogenated hydrocarbon insecticides on cyclic adenosine $3',5'$-monophosphate-^{3}H formation by rat kidney cortex. J Pharmacol Exp Ther 188:265–276.

Kaiser TE, Reichel WL, Locke LN, Cromartie E, Krynitsky AJ, Lamont TG, Mulhern BM, Prouty RM, Stafford CJ, Swineford DM (1980) Organochlorine pesticide, PCB, and PBB residues and necropsy data for bald eagles from 29 states—1975–77. Pestic Monit J 13:145–149.

Kamenov DA, Zolotarev SA (1979) Effect of certain pesticides on ecological and physiological mechanisms of sample groups of wild house mice. Dokl Biol Sci (Engl Transl) 249:1207–1209.

Kan CA, Tuinstra LGM (1976) Accumulation and excretion of certain organochlorine insecticides in broiler breeder hens. J Agric Food Chem 24:775–778.

Keck G, Paubel P, Monneret RJ (1982) Organochlorine and mercury residues in peregrine falcon eggs in France. Bull Environ Contam Toxicol 28:705–709.

Kerr SR, Vass WP (1973) Pesticide residues in aquatic invertebrates. In: Edwards CA (ed) Environmental pollution by pesticides. Plenum Press, London. pp 134–180.

Khaleeq B, Klatt A (1986) Effects of various fungicides and insecticides on emergence of three wheat cultivars. Agron J 78:967–970.

Khan MAQ, Feroz M, Sudershan P (1979) Metabolism of cyclodiene insecticides by fish. In: Khan MAQ, Lech JJ, Menn JJ (eds) Pesticide and xenobiotic metabolism in aquatic organisms. ACS Symp Ser 99:37–56.

King KA, Krynitsky AJ (1986) Population trends, reproductive success, and organochlorine chemical contaminants in waterbirds nesting in Galveston Bay, Texas. Arch Environ Contam Toxicol 15:367–376.

Kinoshita FK, Kempf CK (1970) Quantitative measurement of hepatic microsomal enzyme induction after dietary intake of chlorinated hydrocarbon insecticides. Toxicol Appl Pharmacol 17:288.

Knox JR, Khalifa S, Ivie GW, Casida JE (1973) Characterization of the photoisomers from *cis*- and *trans*-chlordanes, *trans*-nonachlor and heptachlor epoxide. Tetrahedron 29:3869–3879.

Knutson H, Kadoum AM, Hopkins TL, Swoyer GF, Harvey TL (1971) Insecticide useage and residues in a newly developed Great Plains irrigation district. Pestic Monit J 5:17–27.

Krantz WC, Mulhern BM, Bagley GE, Sprunt A IV, Ligas FJ, Robertson WB Jr. (1970) Organochlorine and heavy metal residues in bald eagle eggs. Pestic Monit J 4:136–140.

Kretizer JF (1974) Residues of organochlorine pesticides, mercury, and PCB's in mourning doves from eastern United States—1970–71. Pestic Monit J 7:195–199.

Kruthanut S (1986) Pesticide residues in bats in Thailand. Cour Forschungsinst Senckenberg 87:91–108.

Kulkarni JH, Sardeshpande JS, Bagyaraj DJ (1974) Effect of four soil-applied insecticides on symbiosis of *Rhizobium* sp. with *Arachis hypogaea* Linn. Plant Soil 40:169–172.

Lal R, Saxena DM (1982) Accumulation, metabolism, and effects of organochlorine insecticides on microorganisms. Microbiol Rev 46:95–127.

Lang JT, Rodriguez LL, Livingston JM (1979) Organochlorine pesticide residues in soils from six U.S. Air Force bases, 1975–76. Pestic Monit J 12:230–233.

Lehman AJ (1948) The toxicology of the newer agricultural chemicals. Q Bull Assoc Food Drug Off US 12:82–89.

Lehman AJ (1952) Chemicals in foods: a report to the Association of Food and Drug Officials on current developments. Part II. Pesticides. Section II. Dermal Toxicity. Q Bull Assoc Food Drug Off US 16:3–9.

Le Marchand L, Kolonel LN, Siegel BZ, Dendle WH (1986) Trends in birth defects for a Hawaiian population exposed to heptachlor and for the USA. Arch Environ Hlth 41:145–148.

Leung S, Bulkley RV, Richard JJ (1981) Influence of a new impoundment on pesticide concentrations in warmwater fish, Saylorville Reservoir, Des Moines River, Iowa, 1977–78. Pestic Monit J 15:117–122.

Lichtenstein EP, Mueller CH, Myrdal GR, and Schulz KR (1962) Vertical distribution and persistence of insecticidal residues in soils as influenced by mode of application and a cover crop. J Econ Entomol 55:215-219.

Lichtenstein EP, Schulz KR, Fuhremann TW (1971) Effects of a cover crop versus soil cultivation on the fate and vertical distribution of insecticide residues in soil 7 to 11 years after soil treatment. Pestic Monit J 5:218-222.

Linder RL, Dahlgren RB (1970) Occurrence of organochlorine insecticides in pheasants of South Dakota. Pestic Monit J 3:227-232.

Lissitsky S (1976) Biosynthesis of thyroid hormones. Pharmacol Ther B 2:219-246.

Littrell EE (1986) Shell thickness and organochlorine pesticides in osprey eggs from Eagle Lake, California. Calif Fish Game 72:182-185.

Livingston JM, Jones CR (1981) Living area contamination by chlordane used for termite treatment. Bull Environ Contam Toxicol 27:406-411.

Lockie JD, Ratcliffe DA (1964) Insecticides and Scottish golden eagles. Br Birds 57:89-102.

Longcore JR, Mulhern BM (1973) Organochlorine pesticides and polychlorinated biphenyls in black duck eggs from the United States and Canada—1971. Pestic Monit J 7:62-66.

Louis JB, Kisselbach KC Jr. (1987) Indoor air levels of chlordane and heptachlor following termiticide applications. Bull Environ Contam Toxicol 39:911-918.

Lu P-Y, Metcalf RL, Hirwe AS, Williams JW (1975) Evaluation of environmental distribution and fate of hexachlorocyclopentadiene, chlordene, heptachlor and heptachlor epoxide in a laboratory model ecosystem. J Agric Food Chem 23:967-973.

Luke MA, Matsumoto HT (1986) Pesticide residue analysis in foods. In: Zweig G, Sherma J (eds) Analytical methods for pesticides and plant growth regulators, Vol. XV. Academic Press, Orlando, Fla. pp 161-200.

Lumb GD, Rust JH (1985) The pathological response of the liver and thyroid of the rat to potassium prorenoate (SC-23992). Toxicol Pathol 13:315-324.

Luquet F-M, Goursaud J, Casalis J (1974) La pollution des laits humains français par les residues de pesticides organochlores. L'Alim Vie 62:40-69. (as cited in Mussalo-Rauhammaa et al. 1988)

Macek KJ (1970) Biological magnification of pesticide residues in food chains. Biol Impact Pestic Environ Proc Symp 1:17-21.

Macek KJ, Hutchinson C, Cope OB (1969) Effects of temperature on the susceptibility of bluegills and rainbow trout to selected pesticides. Bull Environ Contam Toxicol 3:174-183.

Macek KJ, Lindberg MA, Sauter S, Buxton KS (1976) Toxicity of four pesticides to water fleas and fathead minnows. Ecological Research Series, US EPA, Duluth, Minnesota. EPA-600/3-76-099.

Macklin AW, Ribelin WE (1971) The relation of pesticides to abortion in dairy cattle. J Am Vet Med Assoc 159:1743-1748.

MacMahon B, Monson RR, Wang HH, Zheng TA (1988) A second follow-up of mortality in a cohort of pesticide applicators. J Occup Med 30:429-432.

Mader WJ (1977) Chemical residues in Arizona Harris' Hawk eggs. Auk 94:587-588.

Magnani B, Powers CD, Wurster CF, O'Connors HB Jr. (1978) Effects of chlordane and heptachlor on the marine dinoflagellate, *Exuviella baltica*, Lohmann. Bull Environ Contam Toxicol 20:1-8.

Mampe CD (1987) Termiticides: Is there a right one? Pest Control 55:26-54.

March RB (1952) The resolution and chemical and biological characterization of some constituents of technical chlordane. J Econ Entomol 45:452-456.

Maronpot RR, Haseman JK, Boorman GA, Eustis SE, Rao GN, Huff JE (1987) Liver lesions in B6C3F1 mice: the National Toxicology Program, Experience and Position. Arch Toxicol Suppl 10:10-26.

Marshall TC, Dorough HW, Swim HE (1976) Screening of pesticides for mutagenic potential using *Salmonella typhimurium* mutants. J Agric Food Chem 24:560-563.

Martin WE (1969) Organochlorine insecticide residues in starlings. Pestic Monit J 3:102-114.

Martin WE, Nickerson PR (1972) Organochlorine residues in starlings—1970. Pestic Monit J 6:33-40.

Matsumura F, Ghiasuddin SM (1983) Evidence for similarities between cyclodiene-type insecticides and picrotoxinin in their action mechanisms. J Environ Sci Hlth Part B 18:1-14.

Matsumura F, Tanaka K (1984) Molecular basis of neuroexcitatory actions of cyclodiene-type insecticides. In: Narahashi TN (ed) Cellular and molecular neurotoxicity. Raven Press, New York. pp 225-240.

McDougall KW, Singh C, Harris CR, Higginson FR (1987) Organochlorine insecticide residues in some agricultural soils on the North Coast Region of New South Wales. Bull Environ Contam Toxicol 39:286-293.

McEwen FL, Stephenson GR (1979) The use and significance of pesticides in the environment. John Wiley and Sons, New York. 538 pp.

McGuire RR, Zabik MJ, Schuetz RD, Flotard RD (1972) Photochemistry of bioactive compounds. Photochemical reactions of heptachlor: kinetics and mechanisms. J Agric Food Chem 20:856-861.

McLane MAR, Stickel LF, Newsom JD (1971) Organochlorine pesticide residues in woodcock, soils, and earthworms in Louisiana, 1965. Pestic Monit J 5:248-250.

McLane MAR, Dustman EH, Clark ER, Hughes DL (1978) Organochlorine insecticide and polychlorinated biphenyl residues in woodcock wings, 1971-72. Pestic Monit J 12:22-25.

McLaughlin J, Marliac J, Verrett MJ, Mutchler MK, Fitzhugh OG (1963) The injection of chemicals into the yolk sac of fertile eggs prior to incubation as a toxicity test. Toxicol Appl Pharmacol 5:760-771.

McNeil EE, Otson R, Miles WF, Rajabalee FJM (1977) Determination of chlorinated pesticides in potable water. J Chromatogr 132:277-286.

Miles JRW, Tu CM, Harris CR (1969) Metabolism of heptachlor and its degradation products by soil microorganisms. J Econ Entomol 62:1334-1338.

Miles JRW, Tu CM, Harris CR (1971) Degradation of heptachlor epoxide and heptachlor by a mixed culture of soil microorganisms. J Econ Entomol 64:839-841.

Miyazaki A, Hotta T, Marumo S, Sakai M (1978) Synthesis, absolute stereochemistry, and biological activity of optically active cyclodiene insecticides. J Agric Food Chem 26:975-977.

Miyazaki A, Sakai M, Marumo S (1980) Synthesis and biological activity of optically active heptachlor, 2-chloroheptachlor and 3-chloroheptachlor. J Agric Food Chem 28:1310-1311.

Montz WE, Scanlon PF, Oderwald RG, Young RW, Kirkpatrick RL, Gwynn JV (1983) Fat-soluble chemical residues in bobwhite quail from Virginia. Va J Sci 34:130.

Moore GL, Greichus YA, Hugghins EJ (1968) Insecticide residues in pronghorn antelope of South Dakota. Bull Environ Contam Toxicol 3:269–273.

Mora MA, Anderson DW, Mount ME (1987) Seasonal variation of body condition and organochlorines in wild ducks from California and Mexico. J Wildl Mgmt 51:132–141.

Morgan EP (1987) Heptachlor. In: Poisindex. Producers: Rocky Mountain Poison and Drug Center, Emergency Information Center, and University of Colorado Health Science Center. Distributor: Micromedex, Denver, CO.

Moriya M, Ohta T, Watanabe K, Miyazawa T, Kato K, Shirasu Y (1983) Further mutagenicity studies on pesticides in bacterial reversion assay systems. Mutat Res 116:185–216.

Morrison ML, Slack RD, Shanley E Jr. (1978) Declines in environmental pollutants in olivaceous cormorant eggs from Texas, 1970–1977. Wilson Bull 90:640–642.

Moubry RJ, Myrdal GR, Jensen HP (1967) Chlorinated hydrocarbon pesticide residues in or on alfalfa grown in soil with a previous history of aldrin and heptachlor application. Pestic Monit J 1(2):13–14.

Mulhern BM, Reichel WL, Locke LN, Lamont TG, Belisle A, Cromartie E, Bagley G, Prouty RM (1970) Organochlorine residues and autopsy data from bald eagles 1966–68. Pestic Monit J 4:141–144.

Mullins DE, Johnsen RE, Starr RI (1971) Persistence of organochlorine insecticide residues in agricultural soils of Colorado. Pestic Monit J 5:268–275.

Mullis KB, Faloona FA (1987) Specific synthesis of DNA *in vitro* via a polymerase-catalyzed chain reaction. Methods Enzymol 155:335–350.

Murphy SD (1986) Toxic effects of pesticides. In: Klaassen CD, Amdur MO, Doull J (eds) Cassarett and Doull's toxicology, The basic science of poisons, 3rd Ed. MacMillan Publishing, New York. pp 519–581.

Murray DAJ (1979) Rapid micro extraction procedure for analyses of trace amounts of organic compounds in water by gas chromatography and comparisons with macro extraction methods. J Chromatogr 177:135–140.

Mussalo-Rauhamaa H, Pyysalo H, Antervo K (1988) Relation between the content of organochlorine compounds in Finnish human milk and characteristics of the mothers. J Toxicol Environ Hlth 25:1–19.

Nalley L, Hoff G, Bigler W, Schneider N (1978) Pesticide levels in the omental fat of urban gray squirrels. Bull Environ Contam Toxicol 19:42–46.

Nash RG (1983a) Comparative volatilization and dissipation rates of several pesticides from soil. J Agric Food Chem 31:210–217.

Nash RG (1983b) Distribution of butylate, heptachlor, lindane, and dieldrin emulsifiable concentrated and butylate microencapsulated formulations in microagroecosystem chambers. J Agric Food Chem 31:1195–1201.

Nash RG (1984) Extraction of pesticides from environmental samples by steam distillation. J Assoc Off Anal Chem 67:199–203.

Nash RG, Beall ML Jr. (1970) Chlorinated hydrocarbon insecticides: root uptake versus vapor contamination of soybean foliage. Science (Washington, DC) 168:1109–1111.

Nash RG, Harris WG (1973) Chlorinated hydrocarbon insecticide residues in crops and soil. J Environ Qual 2:269–273.

Nash RG, Beall ML Jr., Woolson EA (1970) Plant uptake of chlorinated insecticides from soils. Agron J 62:369–372.

National Research Council (1982) An assessment of the health risks of seven pesticides used for termite control. Prepared for US Dept Navy, Washington, DC NTIS #PB 83-136374. (as cited in USPHS 1987)

NCI (1977a) Bioassay of chlordane for possible carcinogenicity. Tech Rep Ser No. 8. US Dept HEW, Public Health Service, NIH, Bethesda, Maryland. 117 pp.

NCI (1977b) Bioassay of heptachlor for possible carcinogenicity. Tech Rep Ser No. 9, US Dept HEW, Public Health Service, NIH, Bethesda, Maryland. 111 pp.

Negherbon WO (1959) Heptachlor. In: Handbook of toxicology, Vol 3: Insecticides. WB Saunders, Philadelphia. pp 366-375.

Nelson BD (1975) The action on cyclodiene pesticides on oxidative phosphorylation in rat liver mitochondria. Biochem Pharmacol 24:1485-1490.

Nelson BD, Williams C (1971) Action of cyclodiene pesticides on oxidative metabolism in the yeast *Saccharomyces cerevisiae*. J Agric Food Chem 19:339-341.

Nickerson PR, Barbehenn KR (1975) Organochlorine residues in starlings, 1972. Pestic Monit J 8:247-254.

Norstrom RJ, Hallet DJ, Sonstegard RA (1978) Coho salmon (*Oncorhynchus kisutch*) and herring gulls (*Larus argentatus*) as indicators of organochlorine contamination in Lake Ontario. J Fish Res Board Can 35:1401-1409.

Ober A, Valdivia M, Santa Maria I (1987) Organochlorine pesticide residues in Chilean fish and shellfish species. Bull Environ Contam Toxicol 38:528-533.

Ohlendorf HM, Swineford DM, Locke LN (1981) Organochlorine residues and mortality of herons. Pestic Monit J 14:125-135.

Oloffs PC, Szeto SY, Webster JM (1971) Translocation of organochlorine pesticide residues from soils into carrots. Can J Plant Sci 51:547-550.

O'Neill PM, Langlois BE (1976) Effect of heptachlor on the growth, viability and respiration of *Staphylococcus aureus*. Bull Environ Contam Toxicol 16:330-338.

Pardini RS, Heiddker JC, Payne B (1971) The effect of some cyclodiene pesticides, benzenehexachloride and toxaphene on mitochondrial electron transport. Bull Environ Contam Toxicol 6:436-444.

Parlar H, Mansour M, Baumann R (1978) Photoreactions of hydroxychlordene in solution, as solids, and on the surface of leaves. J Agric Food Chem 26:1321-1324.

Peach ME, Schaffner JP, Stiles DA (1973) Movement of aldrin and heptachlor residues in a sloping field of sandy loam texture. Can J Soil Sci 53:459-463.

Pearce PA, Peakall DB, Reynolds LM (1979) Shell thinning and residues of organochlorines and mercury in seabird eggs, eastern Canada, 1970-76. Pestic Monit J 13:61-68.

Peirano WB (1980) Heptachlor—maximum acceptable limit in drinking water. Washington DC, US EPA (A criteria document prepared for the World Health Organization). (as cited in WHO 1984b)

Peraino C, Fry RJM, Staffeldt E, Christopher JP (1975) Comparative enhancing effects of phenobarbital, amobarbital, diphenylhydantoin, and dichlorodiphenyltrichloroethane on 2-acetylaminofluorene induced hepatic tumorigenesis in the rat. Cancer Res 35:2884-2890.

Pitot HC, Barsness L, Goldsworthy T, Kitagawa T (1978) Biochemical characterisation of stages of hepatocarcinogenesis after a single dose of diethylnitrosamine. Nature (London) 271:456-457.

Pitot HC, Goldsworthy TL, Moran S, Kennan W, Glauert HP, Maronpot R, Campbell HA (1987) A method to quantitate the relative initiating and promoting potencies of hepatocarcinogenic agents in their dose-response relationships to altered hepatic foci. Carcinogenesis 8:1491-1499.

Plewa MJ, Wagner ED (1981) Germinal cell mutagenesis in specifically designed maize genotypes. Environ Hlth Perspect 37:61-73.

Podowski AA, Banerjee BC, Feroz M, Dudek MA, Willey RL, Khan MAQ (1979) Photolysis of heptachlor and *cis*-chlordane and toxicity of their photoisomers to animals. Arch Environ Contam Toxicol 8:509-518.

Probst GS, McMahon RE, Hill LE, Thompson CZ, Epp JD, Neal SB (1981) Chemically-induced unscheduled DNA synthesis in primary rat hepatocyte cultures: A comparison with bacterial mutagenicity using 218 compounds. Environ Mutagen 3:11-32.

Prouty RM, Reichel WL, Locke LN, Belisle AA, Cromartie E, Kaiser TE, Lamont TG, Mulhern BM, Swineford DM (1977) Residues of organochlorine pesticides and polychlorinated biphenyls and autopsy data for bald eagles, 1973-74. Pestic Monit J 11:134-137.

Puszkin S, Kochwa S (1974) Regulation of neurotransmitter release by a complex of actin with relaxing protein from rat brain synaptosomes. J Biol Chem 249:7711-7714.

Radhaiah V, Girija M, Rao K (1987) Changes in selected biochemical parameters in the kidney and blood of the fish, *Tilapia mossambica* (Peters), exposed to heptachlor. Bull Environ Contam Toxicol 39:1006-1011.

Radomski JL, Davidow B (1953) The metabolite of heptachlor, its estimation, storage and toxicity. J Pharmacol Exp Ther 107:266-272.

Reichel WL, Cromartie E, Lamont TG, Mulhern BM, Prouty RM (1969) Pesticide residues in eagles. Pestic Monit J 3:142-144.

Reidinger RF Jr., Crabtree DG (1974) Organochlorine residues in golden eagles, United States—March 1964-July 1971. Pestic Monit J 8:37-43.

Rennie RJ (1977) Immunofluorescence detection of *Nitrobacter* in soil during NO_2^- oxidation in the presence of exotic chemicals. Microbios Lett 6:19-26.

Reuber MD (1977a) Hepatic vein thrombosis in mice ingesting chlorinated hydrocarbons. Arch Toxicol 38:163-168.

Reuber MD (1977b) Histopathology of carcinomas of the liver in mice ingesting heptachlor or heptachlor epoxide. Exper Cell Biol 45:147-157.

Reuber MD (1987) Carcinogenicity of heptachlor and heptachlor epoxide. J Environ Pathol Toxicol Oncol 7(3):85-114.

Reynolds SH, Stowers SJ, Patterson RM, Maronpot RR, Aaronson SA, Anderson MW (1987) Activated oncogenes in B6C3F1 mouse liver tumors: implications for risk assessment. Science (Washington, DC) 237:1309-1316.

Rihan TI, Mustafa HT, Caldwell G Jr., Frazier J (1978) Chlorinated pesticides and heavy metals in streams and lakes of northern Mississippi water. Bull Environ Contam Toxicol 20:568-572.

Rimkus G, Wolf M (1987) Contamination of game by harmful substances in Schleswig-Holstein. 2. Communication: residues of dieldrin, heptachlor epoxide and other cyclodiene insecticides in liver fat of hares (*Lepus europaeus* L.). Z Lebensm Unters-Forsch 184:308-312.

Ristow D, Conrad B, Wink C, Wink M (1980) Pesticide residues of failed eggs of Eleonora's Falcon *Falco eleonorae* from an Aegean colony. Ibis 122:74-76.

Ritcey WR, Savary G, McCully KA (1972) Organochlorine insecticide residues in human milk, evaporated milk and some milk substitutes in Canada. Can J Public Hlth 63:125-132.

Robel RJ, Stalling CD, Westfahl ME, Kadoum AM (1972) Effects of insecticides on populations of rodents in Kansas–1965-69. Pestic Monit J 6:115–121.

Rodica G, Stefania M (1973) Effects of some insecticides on the bursa of Fabricius in chicks. Arch Exp Veterinaermed 27:723–728.

Rohr U, Konig W, Selenka F (1985) Influence of pesticides on the release of histamine, chemotactic factors and leukotrienes from rat mast cells and human basophils. Zentralbl Bakteriol Mikrobiol Hyg Abt. 1 Orig B 181:469–486.

Ronald K, Frank RJ, Dougan JL, Frank R, Braun HE (1984) Pollutants in harp seals (*Phoca groenlandica*). I. Organochlorines. Sci Total Environ 38:133–152.

Rosen JD, Sutherland DJ, Khan MAQ (1969) Properties of photoisomers of heptachlor and isodrin. J Agric Food Chem 17:404–405.

Rosene W Jr. (1965) Effects of field applications of heptachlor on bobwhite quail and other wild animals. J Wildl Mgmt 29:554–580.

Rozman K (1984) Phase II enzyme induction reduces body burden of heptachlor in rats. Toxicol Lett 20:5–12.

Ruch RJ, Klaunig JE (1986) Tumor promoter inhibition of mouse hepatocyte intercellular communication. In Vitro Cell Dev Biol 22(3):part II,38A.

Saha JG, Stewart WWA (1967) Heptachlor, heptachlor, epoxide, and gamma-chlordane residues in soil and rutabaga after soil and surface treatments with heptachlor. Can J Plant Sci 47:79–88.

Saha JG, Sumner AK (1971) Organochlorine insecticide residues in soil from vegetable farms is Saskatchewan. Pestic Monit J 5:28–31.

Saito I, Kawamura N, Uno K, Hisanaga N, Takeuchi Y, Ono Y, Iwata M, Gotoh M, Okutani H, Matsumoto T, Fukaya Y, Yoshitomi S, Ohno Y (1986) Relationship between chlordane and its metabolites in blood of pest control operators and spraying conditions. Int Arch Occup Environ Hlth 58:91–97.

Sanders HO (1969) Toxicity of pesticides to the crustacean *Gammarus locustris*. Tech Pap Bur Sport Fish Wildl US No. 25.

Sanders HO, Cope OB (1966) Toxicities of several pesticides to two species of cladocerans. Trans Am Fish Soc 95:165–169.

Sanders HO, Cope OB (1968) The relative toxicities of several pesticides to naiads of three species of stoneflies. Limnol Oceanogr 13:112–117.

Sandhu SS, Warren WJ, Nelson P (1978) Pesticidal residue in rural potable water. J Am Water Works Assoc 70:41–45.

Satsmadjis J, Voutsinou-Taliadouri F (1983) *Mytlis galloprovincialis* and *Parapenaeus longirostris* as bioindicators of heavy metal and organochlorine pollution. Mar Biol (Berlin) 76:115–124.

Savage EP, Keefe TJ, Tesari JD, Wheeler HW, Applehans FM, Goes EA, Ford SA (1981) National study of chlorinated hydrocarbon insecticide residues in human milk, USA. Am J Epidemiol 113:413–422.

Scheufler E, Rozman K (1984) Enhanced total body clearance of heptachlor from rats by *trans*-stilbeneoxide. Toxicology 32:93–104.

Schimmel SC, Patrick JM, Forester J (1976) Heptachlor: toxicity to and uptake by several estuarine organisms. J Toxicol Environ Hlth 1:955–965.

Schmitt CJ, Zajicek JL, Ribick MA (1985) National pesticide monitoring program: Residues of organochlorine chemicals in freshwater fish, 1980–81. Arch Environ Contam Toxicol 14:225–260.

Schneeweis JC, Greichus YA, Linder RL (1974) Organochlorine pesticide residue levels in North American timber wolves—1969-71. Pestic Monit J 8:142-143.

Seidensticker JC IV, Reynolds HV III (1971) The nesting, reproductive performance, and chlorinated hydrocarbon residues in the red-tailed hawk and great horned owl in south-central Montana. Wilson Bull 83:408-418.

Seiler JP (1977) Inhibition of testicular DNA synthesis by chemical mutagens and carcinogens. Preliminary results in the validation of a novel short term test. Mutat Res 46:305-310.

Sethunathan N, Yoshida T (1973) Degradation of chlorinated hydrocarbons by *Clostridium* sp. isolated from lindane-amended, flooded soil. Plant Soil 38:663-666.

Shaffi SA (1979) The acute toxicity of heptachlor for freshwater fishes. Toxicol Lett 4:31-37.

Shain SA, Shaeffer JC, Boesel RW (1977) The effect of chronic ingestion of selected pesticides on rat ventral prostate homeostasis. Toxicol Appl Pharmacol 40:115-130.

Shamiyeh NB, Johnson LF (1973) Effect of heptachlor on numbers of bacteria, actinomycetes and fungi in soil. Soil Biol Biochem 5:309-314.

Shaw GG (1984) Organochlorine pesticide and PCB residues in eggs and nestlings of tree swallows, *Tachycineta bicolor*, in central Alberta. Can Field-Nat 98:258-260.

Shellenberger TS, Lei J, Udale B, Newell GW (1966) Comparative toxicity of DDT, dieldrin, and heptachlor to Japanese and bobwhite quail. Toxicol Appl Pharmacol 8:353-354.

Sherma J, Shafik TM (1975) Multiclass, multiresidue analytical method for determining pesticide residues in air. Arch Environ Chem 3:55-71.

Sherman M, Ross E (1961) Acute and subacute toxicity of insecticides to chicks. Toxicol Appl Pharmacol 3:521-533.

Shirasu Y, Moriya M, Kato K, Furuhashi A, Kada T (1976) Mutagenicity screening of pesticides in the microbial system. Mutat Res 40:19-30.

Sierra M, Santiago D (1987) Organochlorine pesticide levels in barn owls collected in León, Spain. Bull Environ Contam Toxicol 38:261-265.

Sierra M, Teran MT, Gallego A, Diez MJ, Santiago D (1987) Organochlorine contamination in three species of diurnal raptors in León, Spain. Bull Environ Contam Toxicol 38:254-260.

Singhal RL, Kacew S (1976) The role of cyclic AMP in chlorinated hydrocarbon-induced toxicity. Fed Proc Fed Am Soc Exp Biol 35:2618-2623.

Smith GS, Rozman KK, Hallford DM, Khan MF, Rankins DL Jr., Hoefler WC, Holcombe DW (1987a) Elimination of ^{14}C-heptachlor from body burdens of lactating ewes into milk and excreta: effects of exogenous growth hormone. J An Sci 65(suppl 1):356 (abstract #388).

Smith GS, Rozman KK, Hallford DM, Khan MF, Rankins DL Jr. (1987b) Evidence of rapid metabolism of heptachlor by sheep. J An Sci 65(suppl 1):356-357(abstract #389).

Smith RJ (1982) Hawaiian milk contamination causes alarm. Science (Washington, DC) 217:137-140.

Smith RM, Cole CF (1970) Chlorinated by hydrocarbon insecticide residues in winter flounder, *Pseudopleuronectes americanus*, from the Weweantic River estuary, Massachusetts. J Fish Res Board Can 27:2374-2380.

Smith SI, Weber CW, Reid BLC (1970) The effect of injection of chlorinated hydrocarbon pesticides in hatchability of eggs. Toxicol Appl Pharmacol 16:179-185.

Solomon J (1979) Sample cleanup and concentration apparatus for the determination of chlorinated hydrocarbon residues in environmental samples. Anal Chem 51:1861–1863.

Somers JD, Goski BC, Barrett MW (1987) Organochlorine residues in northeastern Alberta otters. Bull Environ Contam Toxicol 39:783–790.

Sperling F, Ewenike HKV (1972) Changes in LD_{50} of parathion and heptachlor following turpentine pretreatment. Environ Res 5:164–171.

Springer MA (1980) Pesticide levels, egg and eggshell parameters of great horned owls. Ohio J Sci 80:184–187.

Squires RF, Casida JE, Richardson M, Saederup E (1983) [^{35}S]t-bicyclo-phosphorothionate binds with high affinity to brain-specific sites coupled to gamma-amino-butyric-acid-A and ion recognition sites. Mol Pharmacol 23:326–336.

St. Omer V, Ecobichon DJ (1971) The acute effect of some hydrocarbon insecticides on the acetylcholine content of rat brain. Can J Physiol Pharmacol 49:79–83.

Stanley CW, Post AP (1967) Determination of carbaryl, chlorinated hydrocarbon pesticide, and organophosphate residues in foodstuffs. Chem Abstr 153:A34.

Stanley CW, Barney JE II, Helton MR, Yobs AR (1971) Measurement of atmospheric levels of pesticides. Environ Sci Technol 5:430–435.

Stehr-Green PA, Wohlleb JC, Royce W, Head MT (1988) An evaluation of serum pesticide residue levels and liver function in persons exposed to dairy products contaminated with heptachlor. J Am Med Assoc 259:374–377.

Stendell RC, Cromartie E, Wiemeyer SN, Longcore JR (1977) Organochlorine and mercury residues in canvasback duck eggs, 1972–73. J Wildl Mgmt 41:453–457.

Stevens LF, Collier CW, Woodham DW (1970) Monitoring pesticides in soils from areas of regular, limited, and no pesticide use. Pestic Monit J 4:145–166.

Stewart DKR, Fox CJS (1971) Persistence of organochlorine insecticides and their metabolites in Nova Scotian soils. J Econ Entomol 64:367–371.

Stewart DKR, Chisolm D, Fox CJS (1965) Insecticide residues in potatoes and soil after consecutive soil treatments of aldrin and heptachlor. Can J Plant Sci 45:72–78.

Stickel LF (1973) Pesticide residues in birds and mammals. In: Edwards CA (ed) Environmental pollution by pesticides. Plenum Press, New York. pp 254–313.

Stickel WH, Hayne DW, Stickel LF (1965a) Effects of heptachlor-contaminated earthworms on woodcock. J Wildl Mgmt 29:132–146.

Stickel WH, Dodge WE, Sheldon WG, DeWitt JB, Stickel LF (1965b) Body condition and response to pesticides in woodcock. J Wildl Mgmt 29:147–155.

Stickel LF, Stickel WH, McArthur RD, Hughes DL (1979) Chlordane in birds: a study of lethal residues and loss rates. In: Diechmann WB (ed) Toxicology and occupational medicine. Elsevier, North Holland, New York. pp 387–396.

Stone WB, Okoniewski JC (1983) Organochlorine toxicants in great horned owls from New York, 1981–1982. Northeast Environ Sci 2:1–7.

Stowers SJ, Wiseman RW, Ward JM, Miller EC, Miller JA, Anderson MW, Eva A (1988) Detection of activated proto-oncogenes in N-nitrosodiethylamine-induced liver tumors: a comparison between B6C3F1 mice and Fischer 344 rats. Carcinogenesis 9:271–276.

Strachan WMJ (1985) Organic substances in the rainfall of Lake Superior: 1983. Environ Toxicol Chem 4:677–683.

Strachan WMJ (1988) Toxic contaminants in rainfall in Canada: 1984. Environ Toxicol Chem 7:871–877.

Suns K, Rees GA (1978) Organochlorine contaminant residues in young-of-the-year spottail shiners from Lakes Ontario, Erie, and St. Clair. J Great Lakes Res 4:230–233.

Szaro RC, Coon NC, Kolbe E (1979) Pesticide and PCB of common eider, herring gull and great black-backed gull eggs. Bull Environ Contam Toxicol 22:394–399.

Taira MC, San Martin de Viale LC (1980) Porphyrinogen carboxy-lyase from chick embryo liver—*in vivo* effect of heptachlor and lindane. Int J Biochem 12:1033–1038.

Takahashi W, Saidin D, Takei G, Wong L (1981) Organochlorine pesticide residues in human milk in Hawaii, 1979–1980. Bull Environ Contam Toxicol 27:506–511.

Talekar NS, Kao HT, Chen JS (1983) Persistence of selected insecticides in subtropical soil after repeated biweekly applications over two years. J Econ Entomol 76:711–716.

Tashiro S, Matsumura F (1978) Metabolism of *trans*-nonachlor and related chlordane components in rat and man. Arch Environ Contam Toxicol 7:113–127.

Taylor AW, Glotfelty DE, Glass BL, Freeman HP, Edwards WM (1976) Volatilization of dieldrin and heptachlor from a maize field. J Agric Food Chem 24:625–631.

Taylor AW, Glotfelty DE, Turner BC, Silver RE, Freeman HP, Weiss A (1977) Volatilization of dieldrin and heptachlor residues from field vegation. J Agric Food Chem 25:542–548.

Teeple SM (1977) Reproductive success of herring gulls nesting on Brothers Island, Lake Ontario, in 1973. Can Field-Nat 91:148–157.

Teichman J, Bevenue A, Hylin JW (1978) Separation of polychlorobiphenyls from chlorinated pesticides in sediment and oyster samples for analysis by gas chromatography. J Chromatog 151:155–161.

Telang S, Tong C, Williams GM (1981) Induction of mutagenesis of carcinogenic polycyclic aromatic hydrocarbons but not by organochlorine pesticides in the ARL/HGPRT mutagenesis assay. Environ Mutagen 3:359.

Telang S, Tong C, Williams GM (1982) Epigenetic membrane effects of a possible tumor promoting type on cultured liver cells by the non-genotoxic organochlorine pesticides chlordane and heptachlor. Carcinogenesis 3:1175–1178.

Teran MT, Sierra M (1987) Organochlorine insecticides in trout, *Salmo trutta fario* L., taken from four rivers in León, Spain. Bull Environ Contam Toxicol 38:247–253.

Thomas TC, Seiber JN (1974) Chromosorb 102, an efficient medium for trapping pesticides from air. Bull Environ Contam Toxicol 12:17–25.

Townsend LR, Specht HB (1975) Organophosphorus and organochlorine pesticide residues in soils and uptake by tobacco plants. Can J Plant Sci 55:835–842.

Train RE (1976) Consolidated heptachlor/chlordane hearings. Fed Regis 41:7552–7585.

Trautman WL, Chesters G, Pionke HB (1968) Organochlorine insecticide composition of randomly selected soils from nine states—1967. Pestic Monit J 2:93–96.

Truhlar JF, Reed LA (1976) Occurrence of pesticide residues in four streams draining different land-use areas in Pennsylvania, 1969–71. Pestic Monit J 10:101–110.

Turner BC, Taylor AW, Edwards WM (1972) Dieldrin and heptachlor residues in soybeans. Agron J 64:237–239.

US Department of Agriculture (1969) Monitoring agricultural pesticide residues, 1965–1967. US Dept Agr, Plant Pest Control Div ARS 81-32. (as cited in Stickel 1973)

USPHS (1987) Toxicological profile for heptachlor/heptachlor epoxide (draft). Agency for Toxic Substances and Disease Registry, US Public Health Service, Atlanta, GA. 116 pp.

Vannucchi C, Sivieri S, Ceccanti M (1978) Residues of chlorinated naphthalenes, other hydrocarbons and toxic metals (Hg, Pb, Cd) in tissues of Mediterranean seagulls. Chemosphere 7:483–490.

Vargova M, Kovalcik V (1969) Influence of heptachlor on the effectiveness of aminopyrine and morphine in mice. Toxicol Appl Pharmacol 15:225–228.

Veierov D, Aharonson N (1980) Economic method for analysis of fluid milk for organochlorine residues at the 10 ppb level. J Assoc Off Anal Chem 63:532–535.

Veith GD, Kuehl DW, Leonard EN, Welch K, Pratt G (1981) Polychlorinated biphenyls and other organic chemical residues in fish from major United States watersheds near the Great Lakes, 1978. Pestic Monit J 15:1–8.

Voutsinou-Taliadouri F, Satsmadjis J (1982) Influence of metropolitan waste on the concentration of chlorinated hydrocarbons and metals in striped mullet. Mar Pollut Bull 13:266–269.

Wade AE, Norred WP (1976) Effect of dietary lipid on drug-metabolizing enzymes. Fed Proc Fed Am Soc Exp Biol 35:2475–2479.

Wagstaff DJ, McDowell JR, Paulin HJ (1980) Heptachlor residue accumulation and depletion in broiler chickens. Am J Vet Res 41:765–768.

Waliszewski SM, Szymczynski GA (1985) Inexpensive, precise method for the determination of chlorinated pesticide residues in soil. J Chromatogr 321:480–483.

Wang HH, MacMahon B (1979a) Mortality of pesticide applicators. J Occup Med 21:741–744.

Wang HH, MacMahon B (1979b) Mortality of workers employed in the manufacture of chlordane and heptachlor. J Occup Med 21:745–748.

Wang Y-S, Chiu WT-F, Chang F-P, Chen Y-L (1988) Decline of chlorinated hydrocarbon insecticides residues in the tea-garden soils of Taiwan. Proc Natl Sci Counc Repub China Part B 12:9–13.

Weatherholtz WM, Webb RE (1971) Influence of dietary protein on the activity of microsomal epoxidase in the growing rat. J Nutrition 101:9–12.

Webb GD, Sharp RW, Abell EM (1976) Effects of some insecticides on sodium efflux from human erythrocytes. Fed Proc Fed Am Soc Exp Biol 35:835.

Webb GD, Sharp RW, Feldman JD (1979) Effect of insecticides on the short-circuit current and resistance of isolated frog skin. Pestic Biochem Physiol 10:23–30.

Weber RR, Montone RC (1988) Distribution of organochlorines in the atmosphere of the south Atlantic and Antarctic oceans. Abstr Pap Chem Congr North Am Cont 3:AGRO 94.

Welch RM, Levin W, Kuntzman R, Jacobson M, Connery AH (1971) Effect of halogenated hydrocarbon insecticides on the metabolism and uterotropic action of estrogens in rats and mice. Toxicol Appl Pharmacol 19:234–246.

Weseloh DV, Teeple SM, Gilbertson M (1983) Double-crested cormorants of the Great Lakes: egg-laying parameters, reproductive failure, and contaminant residues in eggs, Lake Huron 1972–1973. Can J Zool 61:427–436.

White A, Handler P, Smith EL, Hill RL, Lehman ER (1978) Principles of Biochemistry, 6th Ed. McGraw-Hill, New York. p 1118.

White DH (1976) Nationwide residues of organochlorines in starlings, 1974. Pestic Monit J 10:10–17.

White DH (1979a) Nationwide residues of organochlorine compounds in starlings (*Sturnus vulgaris*), 1976. Pestic Monit J 12:193–197.

White DH (1979b) Nationwide residues of organochlorine compounds in wings of adult mallards and black ducks, 1976–1977. Pestic Monit J 13:12–16.

White DH, Stendell RC, Mulhern BM (1979) Relations of wintering canvasbacks to environmental pollutants—Chesapeake Bay, Maryland. Wilson Bull 91:279–287.

White DH, King KA, Prouty RM (1980) Significance of organochlorine and heavy metal residues in wintering shorebirds at Corpus Christi, Texas, 1976–1977. Pestic Monit J 14:58–63.
WHO (1984a) Guidelines for drinking water quality. Vol. 2. Health criteria and other supporting information. World Health Organization, Geneva, Switzerland. pp 208–212.
WHO (1984b) Heptachlor. Environmental Health Criteria No. 38. World Health Organization, Geneva, Switzerland. 81 pp.
Wiemeyer SN, Cromartie E (1981) Relationships between brain and carcass organochlorine residues in ospreys. Bull Environ Contam Toxicol 27:499–505.
Wiemeyer SN, Mulhern BM, Ligas FJ, Hensel RJ, Mathisen JE, Robards FC, Postupalsky S (1972) Residues of organochlorine pesticides, polychlorinated biphenyls, and mercury in bald eagle eggs and changes in shell thickness—1969 and 1970. Pestic Monit J 6:50–55.
Wiemeyer SN, Lamont TG, Locke LN (1980) Residues of environmental pollutants and necropsy data for eastern United States ospreys, 1964–1973. Estuaries 3:155–167.
Wiemeyer SN, Schmeling SK, Anderson A (1987) Environmental pollutant and necropsy data for ospreys from the eastern United States, 1975–1982. J Wildl Dis 23:279–291.
Wiersma GB, Tai H, Sand PF (1972a) Pesticide residues in soil from eight cities—1969. Pestic Monit J 6:126–129.
Wiersma GB, Tai H, Sand PF (1972b) Pesticide residue levels in soils, FY 1969—National Soils Monitoring Program. Pestic Monit J 6:194–228.
Wilkinson ATS, Finlayson DG, Morley HV (1964) Toxic residues in soil 9 years after treatment with aldrin and heptachlor. Science (Washington, DC) 143:681–682.
Williams DT, Benoit FM, McNeil EE, Otson T (1978) Organochlorine pesticide levels in Ottawa drinking water, 1976. Pestic Monit J 12:163.
Williams GM (1980) Classification of genotoxic and epigenetic hepatocarcinogens using liver culture assays. Ann NY Acad Sci 349:273–282.
Williams GM (1983) Epigenetic effects of liver tumor promoters and implications for health effects. Environ Hlth Perspect 50:177–183.
Williams GM, Numoto S (1984) Promotion of mouse liver neoplasms by the organochlorine pesticides chlordane and heptachlor in comparison to dichlorodiphenyltrichloroethane. Carcinogenesis 5:1689–1696.
Worthing CR (1979) The pesticide manual, 6th ed. British Crop Protection Council Publications, Croydon, UK. (as cited in WHO 1984b)
Wright BS (1965) Some effects of heptachlor and DDT on New Brunswick woodcocks. J Wildl Mgmt 29:172–185.
Wright CG, Leidy RB (1978) Chlorpyrifos residues in air after application to crevices in rooms. Bull Environ Contam Toxicol 19:340–344.
Wright CG, Leidy RB (1982) Chlordane and heptachlor in the ambient air of houses treated for termites. Bull Environ Contam Toxicol 28:617–623.
Yamaguchi I, Matsumura F, Kadous AA (1979) Inhibition of ATP-ases by heptachlorepoxide in rat brain. Pestic Biochem Physiol 11:285–293.
Yamaguchi I, Matsumura F, Kadous AA (1980) Heptachlor epoxide: Effects on calcium-mediated transmitter release from brain synaptosomes in rat. Biochem Pharmacol 29:1815–1824.
Yap HH, Desaiah D, Cutkomp LK, Koch RB (1975) *In vitro* inhibition of fish brain ATPase activity by cyclodiene insecticides and related compounds. Bull Environ Contam Toxicol 14:163–167.

Yoneyama K, Matsumura F (1981) Reductive metabolism of heptachlor, parathion, 4,4′-dichlorobenzophenone, and carbophenothion by rat liver systems. Pestic Biochem Physiol 15:213–221.

Yotti LP, Chang CC, Trosko JE (1979) Elimination of metabolic cooperation in Chinese hamster cells by a tumor promoter. Science (Washington, DC) 206:1089–1091.

Zavon MR, Tye R, Latorre L (1969) Chlorinated hydrocarbon insecticide content of the neonate. Ann NY Acad Sci 160:196–200.

Zitko V, Choi PMK (1972) PCB and p,p'-DDE in eggs of cormorants, gulls, and ducks from the Bay of Fundy, Canada. Bull Environ Contam Toxicol 7:63–64.

Zubillaga HV, Sericano JL, Pucci AE (1987) Organochlorine pesticide contents of tributaries into Blanca Bay, Argentina. Water Air Soil Pollut 32:43–53.

Manuscript received October 24, 1988; accepted February 2, 1989.

Subject Index

Acceptable daily intake (ADI), heptachlor, 115
Actinomycetes, growth inhibition by heptachlor, 74
Actinon, isotope in actinium series, 5, 7
Acute toxicity, heptachlor, 113
Adenocarcinoma, radon induced, 30
ADI, Heptachlor, 115
Aitken particles, polonium attachment, 5
Alligator, heptachlor residues in meat, 76
Alpha particles, properties, 34
Alpha radiation, from radon, 5
Ames mutagenicity test, 106
Amphibia, heptachlor residues, 78
Amphibians, heptachlor content from soil, 76
Aquatic animals, heptachlor effects, 92 ff.
Arachis hypoqaea, heptachlor effects, 74
Aves, heptachlor residues, 79 ff.

Bacillus subtilis, mutagenicity test, 106
Bacteria, growth inhibition by heptachlor, 74
Bats, heptachlor residues, 89
Becquerels, formerly curies, 9
Beta radiation, from radon, 5
Bioaccumulation, heptachlor, 96
Bioconcentration, heptachlor, 96
Birds
 heptachlor LC_{50} values, 77
 heptachlor residues, 79 ff.
 heptachlor tissue and egg residues, 77
 reproductive effects from pesticides, 77
Bladder cancer, heptachlor applicators, 116
Bq, abbreviation for becquerel, 9
Bubo virginianus, heptachlor effects, 77
Bunsen particles, polonium attachment, 5

Buteo lineatus, heptachlor effects, 77
Butterfat, heptachlor residues, 97

Cancer
 in heptachlor applicators, 116
 skin, radon related, 31
 stomach, radon related, 31
Carcinogenicity, heptachlor, 107, 108
Carcinoma, hepatocellular from heptachlor, 107
Children, radon effects, 28
Chlordene epoxide, 69
Chronic toxicity, heptachlor, 115
Cigarette smoke and radon, interaction, 37
Cigarette smoking and radon exposure, interaction, 25
Clathrates, formation from radon, 3
Corn, residues from soil heptachlor, 76
Cortical excitability, heptachlor effects, 103
Crocodile, heptachlor residues in eggs, 76
Crocodylus acutus, heptachlor residues in eggs, 76
Crustacea, heptachlor LC_{50}s, 93
Curies, also known as becquerels, 9
Cyclodienes, heptachlor, 61 ff
Cytochrome P-450, reduction by heptachlor, 101

Deaths, in heptachlor applicators, 116
Dechlorination, heptachlor, 69, 99
Dieldrin
 in bird food chains, 91
 soybean root absorption, 74
Diels–Alder reaction, heptachlor synthesis, 63
Dominant lethal studies, heptachlor in mice, 107

Dose calculations, in radon studies, 35
Dose equivalent
　in radon studies, 34
　radon estimates, 37
Dosimetry, radon measurement, 33
Drosophila melanogaster, X-linked recessive lethals, 107
Ducks, heptachlor residues, 80

Eagles, heptachlor residues, 81
Earthworms, heptachlor content from soil, 76
ED_{50}, aminopyrine increase by heptachlor, 102
Eggs, bird, heptachlor residues, 79
Endrin, soybean root absorption, 74
Enzyme induction, heptachlor 101, 102
Epidemiology
　heptachlor exposure, 116
　lung cancer in miners, 12, 20, 24
　radon exposure, 12, 20, 24
Epoxidation
　heptachlor, 69
　heptachlor in plants, 75
Epoxides, as mutagens, 106
Escherichia coli, mutagenicity test, 106
Estrone metabolism, heptachlor increases, 105
Exposure
　humans to heptachlor, 113, 115
　population to ionizing radiation, 11

Falco columbarius, heptachlor effects, 77
Falcon, heptachlor residues, 83
Fatalities, due to radon exposure, 47
Fire ant control, heptachlor bird mortality, 91
Fish
　heptachlor effects, 92 ff.
　heptachlor LC_{50}s, 93
Fluorspar miner, lung cancer, 19
Fungi, growth inhibition by heptachlor, 74

GABA, effects of heptachlor, 104
Gamma radiation, from radon, 5

Glutathione conjugation, epoxides in mammals, 99
Grays, formerly rads, 9
Great horned owl, heptachlor effects, 77
Grouse, heptachlor residues, 84
Gulls, heptachlor residues, 84
Gy, abbreviation for gray, 10

Hawaii, heptachlor epoxide in milk, 112
Hawks
　heptachlor effects, 77
　heptachlor residues, 81
Health effect, heptachlor, 61 ff
Heme synthesis, heptachlor inhibition, 105
Hepatic vein thrombosis, heptachlor caused, 102
Hepatocarcinogenesis, initiation vs. promotion, 110
Hepatocellular carcinoma, from heptachlor, 107
Heptachlor
　absorption, distribution, elimination, 100
　acceptable daily intake, 115
　acute toxicity, 113
　aminopyrine ED50 increase, 102
　analysis, 65
　background air concentration, 68
　bioaccumulation, 96
　cancer in applicators, 116
　carcinogenicity, 107
　chemical structure, 64
　chronic feeding studies, carcinogenicity, 108
　chronic human exposure, 115
　cortical excitation, 103
　deaths in applicators, 116
　dechlorination, 69, 99
　detectable levels in treated homes, 117
　ecological toxicology, 61 ff., 74
　effects on aquatic animals, 92 ff.
　effects on immune system, 105
　effects on kidney, 105
　effects on liver, 101, 102
　elimination by mammals, 100
　environmental concentration, 67 ff.
　epidemiology, 116

epoxidation in plants, 75
epoxidation in soil, 69
extraction methods, 65
fire ant control, 91
fish microsomes, 95
genotoxicity, 110
health effects, 61 ff.
heme synthesis inhibition, 105
hexobarbital sleep reduction, 101
history, 61
human dietary exposure, 97
human exposure, 113
human health hazard, 118
hydrolysis, 69
in bird food chains, 91
in drinking water, 73
in rainwater, 73
LC_{50} values in birds, 77
LC_{50}s in aquatic organisms, 93
LD_{50}s in lab animals, 113
liver histology, 109
liver tumor promoter, 110
mammalian metabolism, 98
maximum permissible concentrations, 112
metabolism in fish, 73
metabolites, 69
microbial metabolism, 74
microsomal enzyme induction, 101
mutagenicity, 106
phototransformation, 67
physical properties, 63
physiological properties, 98
reduction of cytochrome P-450, 101
reproductive effects in mammals, 105
reproductive effects on birds, 91
residues in Mammalia wildlife, 90
residues in vertebrate wildlife, 78
soil behavior, 70
soil residue levels, 71
soil residue transfer to root crops, 75
soybean root absorption, 74
synthesis, 63
teratogenicity, 111
toxicology, 113
transfer from air to water, 72
transformation, 67
U.S. usage, 62

Heptachlor epoxide, 69
  effects on egg hatch, 111
  effects on liver, 101, 102
  effects on neurotransmitter release, 104
  in fish, 94
  in Hawaiian milk, 112
  in human milk, 112
  mutagenicity, 106, 108
  placental transfer in humans, 100
  residues in vertebrate wildlife, 78
  teratogenicity, 111
Hexobarbital, sleep reduction by heptachlor, 101
Histopathology, radon lung cancer, 30
Human exposure
  heptachlor, 113
  heptachlor in milk, 97
Hydrolysis, heptachlor, 69

Immune system, effects of heptachlor, 105
International System radiological units, 9
Ionizing radiation
  color TV contribution, 11
  diagnostic X-ray contribution, 11
  sources, 11
  US population exposure, 11
Iron miners, lung cancer, 18
Isotopes, of radon, 3

Kestrel, heptachlor residues, 83
Kidney, effects of heptachlor, 105

LC_{50}s
  heptachlor in aquatic organisms, 93
  heptachlor in birds, 77
  heptachlor in lab animals, 113
Lead, as end product of radiation, 5
Liver, effects of heptachlor, 101, 102
Lung, anatomy, 34
Lung cancer
  age vs. radon exposure, 42
  annual deaths from radon, 47
  facilitators, 29
  in miners, 13, 20 ff.
  initiators, 29

Subject Index

Lung cancer (*cont.*)
  models comparison, 44
  promotors, 29
  radon and smoking interaction, 25 ff.
  radon-caused deaths, 2
  smoking deaths, 2

Mammalia, heptachlor residues, 89 ff.
Mammalian wildlife, heptachlor reproductive effects, 91
Maximum permissible concentrations, heptachlor, 112
Merlin
  heptachlor effects, 77
  heptachlor residues, 83
Metabolism, heptachlor in mammals, 98
Metabolites, heptachlor, 69
Microbial metabolism, heptachlor, 74
Microsomal enzyme, induction by heptachlor, 101
Microsomal oxidation, heptachlor, 99
Microsomes, fish, heptachlor, 95
Milk, heptachlor residues, 97
Miners
  lung cancer, 13
  radon mortality, 13
Modeling, radon risk estimates, 39
Models
  comparison, lung cancer, 44
  radon risk estimates, 38
Mutagenicity, heptachlor, 106

N-nitrosodiethylamine, as carcinogen, 110
Nitrobacter sp., heptachlor effects, 74
Nuclear medicine, radon, 3
Numenium americanus, heptachlor effects, 77

Oat cell carcinoma, radon induced, 30
Osprey, heptachlor residues, 83
Owls, heptachlor residues, 84
Oxidative phosphorylation, heptachlor inhibition, 102

P-450, reduction by heptachlor, 101
Pesticide residues, transfer from soil to plant, 75
Pheasant, heptachlor residues, 84
Phototransformation, heptachlor, 67
Picrotoxin receptors, heptachlor epoxide binding, 105
Placental transfer, heptachlor epoxide in humans, 100

Quail, heptachlor residues, 84

Radiations, from radon, 4
Radioactivity
  naturally occurring series, 5
  units of measurement, 9
Radiological units, International System, 9
Radium-223, decay product half-lives, 8
Radium-224, decay product half-lives, 7
Radium-226, decay product half-lives, 6
Radium-226, radon-222 as decay product, 6
Radon and smoking, interaction, 25, 37
Radon
  animal studies, 37
  as therapeutic agent, 23
  conversion factors, 10
  dose equivalent estimates, 37
  dose equivalent in lung, 34
  dosimetry and measurement, 33
  effects in children, 28
  epidemiological studies, 12
  formation of clathrates, 3
  forms of radiation, 4
  general properties, 3
  health effects, 12
  in indoor air, 1 ff.
  isotope in uranium series, 5, 6
  isotopes, 3
  lung cancer, 2
  lung cancer, age function, 43
  lung cancer histopathology, 30
  lung deposition modeling, 34
  mortality in miners, 13
  nuclear properties, 3
  physical properties, 5

population exposure risk, 41
progeny, cause of lung damage, 5
progeny, smoking lung cancer relationship, 27
risk communication, 47
risk estimates, 38
risk from cumulative exposure, 42
risk modeling, 39
sources, 2
Radon-219 (actinon), decay product of actinium, 8
Radon-219 (actinon), decay product of radium-223, 8
Radon-220 (thoron), decay product of radium-224, 7
Radon-220 (thoron), decay product of thorium, 7
Radon-222, decay product of radium-226, 6
Rads, also known as grays, 9
Rain, heptachlor in, 73
Raptors, organochlorines in food chain, 91
Rems, also known as sieverts, 9
Reproductive effects, heptachlor on birds, 91
Reptiles, heptachlor content from soil, 76
Reptilia, heptachlor residues, 78
Rhizobium sp., heptachlor effects, 74
Risk communication, radon exposure, 47
Risk, due to radon exposure, 47
Risk estimates, radon exposure, 38
Rodentia, heptachlor residues, 90

Saccharomyces cerevisiae, mutagenicity test, 106
Salmonella spp., mutagenicity test, 106
Sieverts, formerly rems, 9
Skin cancer
  heptachlor applicators, 116
  radon related, 31
Smoking and radon exposure, interaction, 25, 37
Snakes, mortality from heptachlor, 77
Soil, heptachlor extraction, 66
Solenopsis saevissima, control with heptachlor, 91

Soybeans, root absorption of heptachlor, 74
Staphylococcus aureus, heptachlor effects, 74
Stomach cancer, radon related, 31
Succinic dehydrogenase, heptachlor inhibition, 102
Sv, abbreviation for sievert, 10

Teratogenicity, heptachlor, 111
Termite control, heptachlor use, 63
Terrestrial vertebrates, heptachlor residues, 78
Thorium radioactive series, illustrated, 7
Thoron (radon-220)
  decay product of radium-224, 7
  decay product of thorium, 7
Thoron, isotope in thorium series, 5, 7
Threshold, radon exposure effects, 32
Tobacco, residues from soil heptachlor, 75
Tobacco, smoking and radon exposure, interaction, 25
Toxicology, heptachlor, 61 ff., 113
Turtles, organochlorine residues, 76

Uranium miners
  Canada, health, 14, 17
  Czechoslovakia, health, 15
  U.S., health, 12
Uranium radioactive series, illustrated, 6

Water, heptachlor extraction, 66
Waterfowl, heptachlor reproductive effects, 91
Wildlife
  heptachlor residues, 78
  mammalian, heptachlor reproductive effects, 91

Yeast, growth inhibition by heptachlor, 74

Zea mays, reverse mutations, heptachlor, 106